T0321368

HUTCHINSON

Trends in Science

PHYSICS

HUTCHINSON
Trends in Science
PHYSICS

Overview by
Chris Cooper

FITZROY DEARBORN PUBLISHERS
CHICAGO · LONDON

Published in the United States of America by
Fitzroy Dearborn Publishers
919 North Michigan Avenue
Chicago, Illinois 60611
OR
Fitzroy Dearborn Publishers
310 Regent Street
London W1B 3AX

First published in the USA and in the UK 2001

ISBN: 1–57958–358–X

*Library of Congress and British Library Cataloguing in
Publication Data are available*

Typeset by
Florence Production Ltd, Stoodleigh, Devon
Printed and bound in Great Britain by
Clays Ltd, Bungay, Suffolk

Contributor
Chris Cooper, writer
and editor specializing
in the physical
sciences

Editorial Director
Hilary McGlynn

Managing Editor
Elena Softley

Project Editor
Heather Slade

Editor
Catherine Thompson

Technical Editor
Rachel Margolis

Production Manager
John Normansell

Production Controller
Stacey Penny

Picture Researcher
Sophie Evans

Contents

Preface   *ix*

Part One

Overview   1

Chronology   25

Biographical Sketches   49

Part Two

Directory of Organizations and Institutions   103

Selected Works for Further Reading   121

Web Sites   127

Glossary   145

Appendix   *261*

Index   *269*

Preface

A physicist of the year 1900 transported to the year 2001 would find the subject transformed out of all recognition. At the beginning of the 20th century he – the physicists of 1900 were almost without exception men – would have worked, wearing a jacket and a stiff collar, with apparatus that probably fitted on a benchtop and was constructed with his own hands or the aid of one or two technicians. His peers were to be found in a few élite institutions of Europe and North America. Very little mathematics was needed for his work. Research was funded largely by the charitable endowments of wealthy patrons of science, past and present; governments concerned themselves little with physics or any other kind of scientific research. Physics was cloistered, a retreat from the turbulence of society and politics. Yet this activity of gifted amateurs with homemade equipment had already penetrated deep into the secrets of matter and was poised to enter the atom itself.

Physics at the beginning of the 21st century could hardly be more different. Its practitioners are men and women of every nation, and their laboratories are to be found in every continent. They often work with multi-million-dollar instruments that are themselves the product of years of research and development in universities and private companies. The teams of researchers in some experiments run to hundreds. They are supported by large and costly teams of engineers, but nonetheless often have to get their own hands dirty. Their work is intensely interesting to government, and many of them work in secrecy in military establishments. Even those who do not are likely to be educated and employed at the taxpayer's expense.

A torrent of information is poured out by the huge physics industry. The visitor from 1900 would find that the best physicists of today understand only a part of the field. Mathematics is now all-pervading, yet so abstruse that theoreticians and experimentalists have separated into two groups, whose members trade their results with each other but rarely cross the frontier between them.

The most profound difference of all would be the change in the spirit and expectations of the science. The foundations of the science of 1900, regarded then as unshakeable, were the principles of mechanics and electromagnetism. Our visitor from the past would find these still being taught. But in first-year university courses, and perhaps earlier, he would find students being taught to put 'classical' physics behind them as a naive approximation to the wider truths of today's physics. But the foundations of modern physics are incomprehensible without mathematics. Millions today read popular science voraciously; but real understanding of the fundamentals is denied to those who do not pursue it mathematically – and that includes many working physicists.

These abstract theories are proposed and discarded with reckless abandon. Textbooks go out of date in 15 years. Modern physics is extraordinarily ambitious, undertaking to explain the very origins and structure of the universe, which the physics of 1900 did not. The physics left behind by the visitor from 1900, though faced with difficulties, was seen as engaged in completing a building that rested on secure foundations. The physics that begins the 21st century is irretrievably committed to perpetual revolution.

Chris Cooper
April 2001

Part One

Overview   1

Chronology   25

Biographical Sketches   49

1   Overview

1900 crisis in physics

Throughout the 19th century the science of physics moved from success to success. By the end of the century the edifice seemed to be essentially complete. The universe obeyed the laws of gravitation and mechanics that the English physicist Isaac Newton (1642–1727) had discovered in the 17th century. Intrinsic to every body was its mass: the greater its mass, the more sluggishly it reacted to an imposed force, and also the stronger the force of gravitation with which it acted on every other body in the universe. Gravitation held the universe together on the astronomical scale.

Another significant group of phenomena, electricity and magnetism, was described by the elegant equations of the Scottish theorist James Clerk Maxwell (1831–1879). It had been known since the work of his compatriot, the experimentalist Michael Faraday (1791–1867), earlier in the 19th century, that electricity and magnetism are linked: an electric current creates a magnetic field, while changes in a magnetic field set up electric currents. In the 1860s and 1870s Maxwell showed how a rapid vibration of an electrical charge or a magnet would cause a disturbance in the electrical and magnetic fields to spread as a wave, travelling at the speed of light. A German physicist, Heinrich Hertz (1857–1894), detected these waves in 1887, created by a repetitive electric spark a few metres away. Maxwell's work made it clear that light was an electromagnetic wave, and that there was an infinite range of electromagnetic radiations with wavelengths longer or shorter than those of visible light.

The two domains of gravitation and electromagnetism were separate. But the concept of energy provided a partial link between them. The energy of a system is its ability to do work. A moving bullet possesses kinetic energy, or energy of motion, because it can splinter wood in being brought to a stop. A mass of steam possesses heat energy because it can drive a steam engine. Coal possesses chemical energy because it can be burned to generate the heat to turn water into steam. Electromagnetic radiation possesses energy because it can warm matter. In turn, heat energy or electrical energy is used up in generating electromagnetic radiation. Energy is like a currency passing through physical processes, never altering in quantity and prescribing exactly how much of one kind of process (say, a chemical reaction) can be exchanged for how much of another (say, an amount of heat). The law of conservation of energy – that there is no change in the total energy in the universe – was taken to be unchallengeable.

There was progress, too, in describing the structure of matter. Although some serious scientists still denied the reality of atoms, they were a minority. The chemical reactions could be explained in terms of the linking up and separating of atoms. Much of the behaviour of gases could be explained by assuming that they were crowds of billiard-ball-like atoms bouncing off the walls of their container. The size, mass, and numbers of atoms could be calculated from careful physical and chemical measurements. However, there were some huge gaps in the edifice of physics waiting to be closed. For example, physicists had no idea what caused the chemical differences between atoms.

In fact, though it was barely realized at the time, science had gained its first glimpse inside the atom with the studies of cathode rays made in 1897 by the English physicist J J Thomson (1856–1840). These rays were seen to stream through low-pressure gas in a glass vessel when a high electrical potential difference was applied between two terminals mounted inside the vessel. Thomson bent the path of the rays with electric and magnetic fields and showed that they were made up of 'particles of electricity'. The mass of these 'particles of electricity' was 1/1836 the mass of the nucleus of the lightest element, hydrogen.

Since these 'electrons' were identical, whatever their source, Thomson suggested that they are components of all atoms, and that ordinary electrical currents are streams of electrons in movement. This was just one step towards explaining the properties of the atoms of the scores of known elements.

But crises were developing in the theory of electromagnetism. Light and other radiations did not behave as they should. Delicate experiments by the German-born US physicist Albert A Michelson (1852–1931) and US physicist Edward W Morley (1838–1923) failed to reveal any effect on light rays of the movement of the Earth around the Sun. The analogy here is to think of a river 5 m/16 ft wide. A boat that travels from one bank to the other and back, at a steady speed relative to the river, travels 10 m/33 ft. A boat that travels 5 m downstream and 5 m back upstream, at the same steady speed relative to the river, also travels 10 m. However the two trip times are measurably different. At the time it was believed that light waves travelled through 'the ether'. Furthermore, the speed of movement of a light source, such as a star, had no discernible effect on the speed of light. There were experiments by others, too, that had similarly puzzling results.

Another problem was to explain the spectrum – the pattern of wavelengths – that is emitted by hot bodies. At low temperatures most of this radiation is in the infrared; at the temperature of the Sun, most is emitted in the visible spectrum; for much higher temperatures, the peak is in the ultraviolet. Electromagnetic theory could not explain this. But few physicists thought in 1900 that this and the other problems of their science would need to be solved by revolution.

Time, space, and gravity

In 1905 a young German-born physicist who had taken Swiss nationality, Albert Einstein (1879–1955), published three scientific papers. One concerned Brownian motion, the dance of tiny particles such as pollen grains in a liquid, visible through a microscope. Einstein explained this as being due to the bombardment of the particles by the ceaseless motion of the molecules of the liquid, and his work made it possible to calculate how many molecules there are in a given mass of the liquid. This work was of major importance, but his other papers were epoch-making. One concerned electromagnetic radiation, and is dealt with in the next section. The third revolutionized our conceptions of space and time themselves.

Einstein was unaware of the experimental results of Michelson and Morley but started his work from Maxwell's equations. These equations, named after Scottish physicist James Clerk Maxwell, showed that oscillating

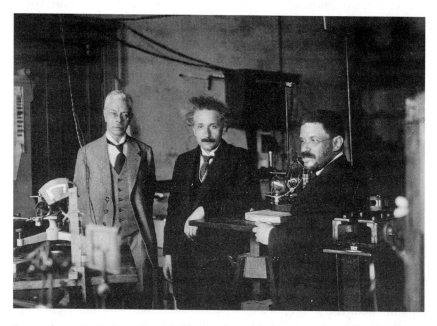

German-born US physicist Albert Einstein with Austrian physicist Paul Ehrenfest (left) and Dutch physicist Pieter Zeeman (right) in Zeeman's laboratory in Amsterdam. University of Michigan IST Willow Run Center Photography Laboratory, courtesy AIP Emilio Segrè Visual Archives, W F Collection

electric charges should produce what are now called electromagnetic waves with a speed similar to that found for light. Einstein made the apparently impossible assumption that the speed of light is the same for all observers, whatever their movement and whatever the movement of the light source. This technically applies only to light in a vacuum; light is slowed down when travelling through a medium such as air or water.

It followed from this that time and space must be different for different observers. Einstein worked out just how they differed. An observer on the ground, John, regards processes in, say, a fast-moving spacecraft as running slow. Jane, in the spacecraft, finds the same about processes on the ground – they are running slow relative to her. In addition, each finds all lengths and distances in the other system to be contracted in the direction of movement.

Einstein calculated the degree of contraction and of 'time dilation' that must occur. For two observers moving at the huge speed of 30,000,000 m per second, one tenth of the speed of light, with respect to each other, the rates of clocks and the lengths of measuring rods are altered by just half of one percent. But at 99% of the speed of light, lengths in the direction of movement are shrunk to one seventh of their original length, and all processes are slowed to one seventh.

These ideas could explain why the Michelson–Morley experiment showed no result. Relative to an observer beyond the Earth and not sharing its motion, the experimental apparatus, everything else on Earth, and the Earth itself, are all shortened in the direction of the Earth's movement. Detailed analysis shows that this just compensates for the effects on apparent speed of the light rays that would be expected classically.

Other results followed. The apparent mass of an object is no longer a constant, as in Newtonian physics, but increases with speed. The amount of energy required to boost it to the speed of light is infinite, and so this is a limiting speed that no physical object can attain. Einstein was able to show that, simply put, mass and energy were equivalent. His most famous, yet often misquoted and misunderstand, equation $E = mc^2$, shows that changing the energy of an object will result in a change in its mass.

Einstein's special theory of relativity, as this work came to be called, was so convincing that physicists around the world took it up immediately. Mathematical analysis by the Hungarian-born US physicist Eugene Wigner (1902–1995) brought out how closely space and time are interlinked in relativity theory, and the word 'space-time' was coined to describe this intimate combination. The picture of a three-dimensional universe moving through time gave way to the picture of a four-dimensional space-time network of events, which different observers analyse in different ways into three dimensions of space and one of time. Einstein quickly took a further giant stride, in his 'general theory' of relativity, published during World War I.

The general theory tackled gravitation. It viewed a body with mass, such as the Sun, as affecting space and time nearby. For example, relative to an observer on the Earth, clocks run slow in the strong gravity field of the Sun. The distortion of space-time also causes bodies with mass to tend to move towards each other. In the weak gravity fields that prevail over the large-scale universe they move in just the ways that Newton's theory predicts to a high degree of approximation. But differences arise for strong gravity fields. Relativity predicts that Mercury's orbit will differ from the one predicted by Newtonian theory – an effect known long before Einstein's theory.

Einstein's general theory also predicts that light rays will be bent by a gravitational field. This prediction of the theory was confirmed by expeditions led by the English astronomers Andrew Crommelin (1865–1939) and Arthur Eddington (1882–1944). Measurements were taken from Sobral in northern Brazil and the island of Príncipe in the Gulf of Guinea on 29 May 1919. Einstein predicted a deflection of 1.74 seconds of arc. The results from Sobral gave the deflection as 1.98 ± 0.16 and those from Príncipe as 1.61 ± 0.40 seconds of arc. More experimental confirmation of relativity has followed in abundance. Time dilation can be seen in today's particle accelerators. Most subatomic particles 'decay' into others in fractions of a second. Their lifetimes are seen to be greatly extended at the near-light speeds to which they are accelerated: their 'onboard clocks' are running slower relative to us.

The brightest objects we know of are quasars, billions of light years from the Earth. They are powered by stars and interstellar matter falling into the powerful gravity field of a massive superdense object. A substantial part of the mass of the victims is turned into energy – in accordance with Einstein's ideas.

The bending of light is dramatically displayed in gravitational lensing, which causes multiple images of some quasars, when a massive object, such as a cluster of galaxies, intervenes. In 1979 the first double image of a quasar, due to gravitational lensing, was observed.

Radiation

In 1900 the German physicist Max Planck (1858–1947) suggested that electromagnetic radiation could not be given out or absorbed in arbitrary amounts but only in separate 'packets', called quanta, with energy proportional to frequency. Ultraviolet radiation had high frequency and could therefore be emitted and absorbed only in high-energy quanta. Infrared radiation had low frequency and therefore low-energy quanta. Visible-light quanta were intermediate in energy. This reduced the amount of radiation given out at higher frequencies because larger amounts of energy were needed to supply the higher-frequency quanta. Planck derived a formula specifying how much energy is radiated at each wavelength by a body for any temperature – that is, its spectrum.

Einstein took this idea forward in 1905, the same year in which he published his special theory of relativity. He wanted to explain some puzzling behaviour of the newly discovered electrons. When light was shone onto certain metals, electrons were ejected. If the light was reduced in intensity, the number of electrons emitted decreased, not surprisingly. What was surprising was that the few electrons ejected by a faint light each had just as much energy as the many ejected by a bright light. Einstein accounted for this by putting to one side a century of experiment showing that light is a type of wave motion and proposing that it was behaving in these circumstances like a stream of bullets. A faint ray of light consists of fewer 'bullets' – but individually they have just as much energy and can knock an electron out of the metal with just as much energy.

These 'light particles' came to be called 'photons'. The wave and particle aspects were linked by the fact that the energy of a photon was proportional to its frequency, as in Planck's theory. How light could simultaneously be a wave, in which energy is spread out over a large volume of space, and a stream of particles, in which the energy is gathered into one place, was incomprehensible. But for the moment, Einstein's picture was accepted as accounting for the photoelectric effect.

In 1923 the US physicist Arthur Holly Compton (1892–1962) demonstrated the particle nature of X-rays, which had been discovered by the German physicist Wilhelm Röntgen (1845–1923) in 1895. X-rays have very short wavelengths and so their photons have high energy. Compton bounced X-ray photons off a stream of electrons. The scattered photons lost some energy, and so had longer wavelengths than the initial X-rays. The paths of the electrons struck by photons were altered. This effect now carries the name of its discoverer, the Compton effect.

Applications

The full gamut of electromagnetic radiations was explored as the 20th century passed. Atoms cannot be seen in even the most powerful visible-light microscopes because light wavelengths are too large. But X-ray wavelengths are of the same order as the spacing between atoms, about one nanometre. Two English physicists, Henry Bragg (1862–1942) and his son Lawrence (1890–1971), bounced X-rays off crystals and obtained complex patterns of bright and dark spots that could be deciphered to reveal the spacings and positions of atoms in the crystal surface. This process became known as X-ray crystallography and led to many discoveries including, in 1952, the structure of DNA by English biophysicist Rosalind Franklin (1920–1958).

Ultraviolet (UV) radiation has wavelengths ranging from about 13 nanometres (1 nanometre (1nm) is 10^{-9} m or 0.00000001 m) up to about 400 nanometres, the border of visible light. The astronomical sky can now be studied at UV wavelengths with detectors in satellites, flown above the atmosphere which absorbs much of the UV.

Beyond the other end of the visible spectrum, astronomers study the sky at infrared (IR) wavelengths. Detectors have to be attached to telescopes on mountain tops or flown on satellites to avoid the IR-absorption of the atmosphere. The detectors themselves have to be chilled to reduce their own IR radiation. IR detectors are used in hosts of everyday devices, from TV remote controls to thermal imagers for detecting people buried under snow or debris.

Radio waves were used first for wireless telegraphy, first demonstrated by the Italian electrical engineer Guglielmo Marconi in the 1890s. A number of engineers then developed means of enabling radio waves to carry speech and music. In 1920 the first true radio station began broadcasting in Pittsburgh, Pennsylvania, USA. By 1931 Russian-born US electronics engineer Vladimir Zworykin (1889–1981) at the RCA corporation had developed the fully electronic television system, in which an electron beam 'paints' the picture on the inside of a cathode ray tube. During World War II radio waves of very short wavelengths were developed to make radar possible.

Even the use of visible light was transformed. Lasers were developed in the late 1950s by US physicist Theodore Maiman (1927–). Lasers generate light of unprecedented coherence and intensity. They are used for count-less tasks requiring precision measurement, from finding the distance from the Earth to the Moon to measuring the profiles of car parts very accu-rately. Laser light is indispensable to making the three-dimensional images called holograms, which were invented in 1947 by Hungarian-born British physicist Dennis Gabor (1900–1979).

Deeper into the atom

Thomson's discovery of the electron had been the first glimpse into the interior of the atom. The next came with a chance discovery by French physicist Henri Becquerel (1852–1908), in 1896. Becquerel found that compounds containing uranium gave out invisible radiations that could affect photographic film, even when the film was wrapped in light-excluding paper. This new radiation was soon found to be a mixture of three types of radiation. One, called alpha radiation, could be blocked by aluminium foil only a few thousandths of a centimetre thick and consisted of a stream of positively charged particles. These positively charged parti-cles were later found to consist of two neutrons and two protons, making them nuclei of a helium atom. Another type, beta radiation, consisted of a stream of negatively charged particles, which were as penetrating as X-rays. Becquerel showed that these beta particles were fast-moving electrons. Whilst beta particles are indeed 'fast moving electrons', to be a beta particle, the particle really needs to originate in the nucleus of an atom.

The third type of radioactivity was more penetrating than X-rays and was not deflected by electric or magnetic fields. Named gamma radiation, it turned out to consist of electromagnetic radiation of a wavelength even shorter than that of X-rays. Gamma radiation has found a role in medicine killing micro-organisms and sterilizing instruments. Physicists and chemists worked to establish which substances in nature were radioactive and which were not. The Polish-born French physicist Marie Curie (1867–1934), assisted by her French husband Pierre Curie (1859–1906), separated a few grammes of highly radioactive material from several tonnes of the uranium ore pitchblende. In this tiny sample she found two new elements, which she named polonium and radium. Radium proved to be millions of times more radioactive than uranium.

The radioactivity of an element did not depend at all on the substances with which that element was combined, or on the surrounding conditions of temperature or pressure. Physicists suspected that it was a process originating within individual atoms and therefore unaffected by other atoms.

Understanding atomic structure

The Cavendish Laboratory in Cambridge University was headed by a New Zealand-born physicist, Ernest Rutherford (1871–1937), one of the most brilliant experimentalists of all time. In 1907 two of his students, Hans Geiger (1882–1945), a German, and Ernest Marsden (1889–1970), a New Zealander, observed alpha particles passing through thin metal films. When they emerged, the alpha particles struck a plate coated with phosphorescent zinc sulphide, emitting tiny flashes of light that the experimenters painstakingly counted.

Geiger and Marsden observed slight deflections of the electrons' paths, averaging less than one degree. Rutherford suggested that Marsden look to see whether any of the alpha particles were deflected backwards – that is, through an angle of 90° or greater. Marsden found that a tiny proportion – about one in 20,000 – were indeed bounced back by something they encountered in the gold foil. Rutherford said, 'It was almost as incredible as if you fired a 15-inch shell at a piece of tissue paper and it came back and hit you.'

Rutherford realized that the alpha particles were being strongly repelled by something that had a relatively large mass and carried a concentration of positive charge. The positive charge and mass of the atom must be concentrated into a small region at the centre of the atom, which he called the nucleus.

At the University of Manchester, UK, another English physicist, Henry Moseley (1887–1915), studied the X-rays given out by atoms. He used a new theory of the structure of the atom put forward by a Danish student of Rutherford, Niels Bohr (1885–1962), which is described in the next section.

New Zealand-born British physicist Ernest Rutherford (left) showing English physicist J J Thomson to his seat for the Cavendish Laboratory annual photograph. AIP Emilio Segrè Visual Archives, Bainbridge Collection

Moseley was able to calculate the nuclear charge very accurately and found that, within the limits of experimental error, it was always a whole-number multiple of the charge on the electron. This number was equal to the atom's atomic number – its position in the periodic table. Thus titanium, element number 22, was measured to have a positive nuclear charge of 21.99 units.

But how was the nucleus constructed? In 1919 Rutherford reported experiments in which high-speed alpha particles from a radioactive source collided with nitrogen atoms in the air and knocked out particles with a single positive charge. These particles were nuclei of hydrogen atoms and so had a mass approximately equal to a unit atomic weight. Later Rutherford dubbed these particles 'protons'. The proton seemed to be a building block of the nucleus. Usually in these experiments the nitrogen nucleus absorbed an alpha particle, and then lost a proton while itself turning into an oxygen nucleus. This was the first experiment in which an atom of one element was artificially converted into an atom of another element.

Another Cavendish physicist, James Chadwick (1891–1974), studied a mysterious penetrating radiation that certain elements, such as beryllium, gave out after being bombarded with fast alpha particles. He found that the radiation consisted of electrically uncharged particles with slightly more mass than the proton. Further experiment showed that the neutron is a fundamental particle, just as basic as the proton. The nucleus of the atom consists of protons and neutrons – the nucleus of the nitrogen atom, for example, contains seven protons and seven neutrons.

The neutron is not stable: outside the nucleus it disintegrates, turning into a proton and emitting an electron. In 1933 the Italian theorist Enrico Fermi (1901–1954), explained how this is possible. Both the proton and the electron come into existence at the moment when the neutron ceases to exist – just as a photon comes into existence at the moment when an atom emits light. In the subatomic world the equivalence of mass and energy is observed and experimental evidence has never failed to support the view presented by Einstein.

Quantum physics

The new discoveries brought a new crisis in physics. How could the solar-system model atom survive for more than a fraction of a second? Just as the oscillating electrons in a radio transmitter's antenna give off radio waves, so the whirling electrons in the atom should have given off electromagnetic radiation and fallen into the nucleus.

A young Danish physicist, Niels Bohr, produced a theory based on this fact while he was studying under Rutherford, who had moved to the University of Manchester, UK. Bohr assumed that only those orbits are possible for which the energy is a whole-number multiple of a basic energy. Once again the idea of the quantum of energy was appearing in physics. On this assumption Bohr was able to calculate the permitted orbits in the simplest atom, that of hydrogen. He found the innermost, lowest-energy orbit to have a diameter of 10^{-12} m, in good agreement with the estimated size of the hydrogen atom.

Bohr pictured the electron as being able to jump upwards from one orbit to another of higher energy, farther out, by absorbing the energy of a photon of just the right frequency. The electron could fall from outer orbits to inner ones, giving out energy in the form of electromagnetic waves. Bohr could calculate the wavelengths of light given out and absorbed by hydrogen. Details of the hydrogen spectrum had been put on a mathematical footing by the Swiss school teacher Johann Balmer (1825–1898), who, in 1855, discovered a simple mathematical formula that gave the spectral lines. This became known as the Balmer series.

The results obtained by Bohr were found to be in agreement with the Balmer series. In 1924 a French student of physics, Louis de Broglie (1892–1987), suggested in his doctoral thesis that associated with the electron and all other subatomic particles are pilot waves, their wavelengths related to the energy of the particle. The permitted orbits for the electrons in an atom were those where a whole number of wavelengths fitted precisely into the circumference of the orbit. The US physicists Clinton Davisson (1881–1958) and Lester Germer (1896–1971) demonstrated that electrons behave like waves when they are diffracted by crystals, in the same way that X-rays had been found to do.

In 1926 the German physicist Erwin Schrödinger (1887–1961) derived an equation that was to be the key to the newly emerging physics. It describes the motion of any particle as a wave motion, in accordance with de Broglie's ideas, and successfully solved certain problems. In 1928 the British physicist Paul Dirac (1902–1984) applied relativity theory to the Schrödinger wave function and found that electrons have a property called spin. When an electric charge moves in a circle, it creates a magnetic field. Electrons in an atom do this, so that atoms are like tiny magnets. But individual electrons also behave like tiny magnets. This can be thought of as due to the electrons rotating, although it became increasingly clear that this kind of picture taken from the large-scale world could be misleading.

Antimatter and uncertainty

Dirac went on to show that the electron must have a counterpart particle. At first it was thought that the proton must be this counterpart, but later it was realized that there must be an anti-electron, with the same mass but all its other properties – charge, spin, and other properties later discovered – opposite to those of the electron. This particle was observed in 1932 by the US physicist Carl Anderson (1905–1991) and named the positron (for positive electron). Positrons are commonly produced in radioactivity, but are destroyed as soon as they encounter an ordinary electron. Every particle has its antiparticle: the negatively charged antiproton was discovered in 1955, the antineutron in 1956.

The new quantum physics had many mysterious features. Werner Heisenberg (1901–1976), a German physicist, stated the uncertainty principle, or indeterminacy principle, according to which it is impossible to determine certain properties of particles precisely and simultaneously. For example, the position and momentum of an electron at a given moment cannot be measured precisely. You could observe an electron by firing X-rays at it, for instance, and observing the scattered X-rays, as Compton had done. Heisenberg analysed the disturbance of the electron by its collision with the X-ray, and showed that if the experiment was set up to measure velocity accurately, the momentum of the electron would be highly uncertain; if instead the position was measured accurately, the momentum would be uncertain.

Nineteenth-century physics had been deterministic – knowledge of the state of any system would allow its future to be predicted. The more accurately you measured its properties, the farther into the future you could predict what would happen in that system. The new quantum physics, by contrast, was probabilistic. For example, while it was possible to say that a radioactive atom had a 50% chance of decaying in, say, a year, it was fundamentally impossible to make any more definite prediction about when it would decay.

Niels Bohr argued that it was wrong to regard quantum indeterminacy as a limitation on what we can discover about some underlying realities; it was meaningless to talk of an electron having a definite position and velocity. Among other things, this meant that Bohr's own early picture of the atom as containing electrons moving with definite speeds in definite orbits had to be abandoned.

Similarly, it is meaningless to ask whether the electron is a particle or a wave: in some circumstances it behaves like one, in others it behaves like the other.

Bohr's view has been generally accepted by physicists. Quantum physics grew even stranger in the decades following its birth in the 1920s and 1930s. The British theoretician John Bell (1928–1990) did important work in 1964 that showed how systems that have interacted with each other – for example, particles created in a subatomic process – remain 'entangled' when they separate, even if they travel light years apart. A major research effort is currently in progress to explore these effects.

The framework of matter

Quantum theory was highly successful in explaining chemical forces between atoms, despite the fact that the Schrödinger wave function could be solved only for a very few simple cases, such as the hydrogen atom. For multi-electron atoms, only approximate solutions could be found. These represented electrons inside the atoms as 'electron clouds', called orbitals.

The density of an orbital showed the probability that an electron would be found there. A number of orbitals formed a 'shell', containing electrons with roughly the same energy. In a normal atom electrons occupied the lowest available shells. The innermost shell could hold two electrons, the second eight, the third 18, and so on. Within the shells electrons can occupy slightly different energy levels.

Increasingly, over the years since quantum mechanics was first developed in the 1920s and 1930s, experimental work and theoretical progress have mapped out the energy levels that the electrons occupy in the atoms of the 92 naturally occurring elements. The way in which atoms combine and redistribute their electrons has been explained.

As well as chemical reactions, the chemical properties of materials have been further and further explained. Metals, for example, owe their characteristic properties to the fact that outermost electrons in the atoms become separated and form a 'sea' of electrons in the materials. These electrons can flow easily, making it easy for electric currents to flow through the metal and also for heat to be conducted through it.

Physicists studied matter in extreme conditions. At high temperatures, all substances form gases, in which atoms rush around separately. At higher temperatures still, the electrons are torn from the atoms in collisions. A fluid called a plasma is formed, consisting of negatively charged electrons mixed with positively charged ions (atoms that have lost some or all of their electrons). The motion of these charged particles creates magnetic fields that in turn affect the motion of the particles, so that the plasma's behaviour is very different from that of an ordinary gas. The Swedish physicist Hannes Alfvén (1908–1995) analysed and explained the behaviour of plasmas, which are important in nature. The air is turned into a plasma in lightning strikes and the Sun and stars consist of plasma.

Physicists also found that matter behaves strangely at low temperatures. In 1908 the Dutch physicist Heike Kamerlingh Onnes (1853–1926) liquefied helium at 4.2 K (–268.9°C/–452.0°F). A few years later he found that mercury became superconducting when cooled to such temperatures. It lost its electrical resistance completely, and a current could be set up in it that would circulate for ever if left undisturbed, without further energy needing to be put in. The explanation for this was provided by the BCS theory, named from the initials of the US physicists – John Bardeen (1908–1991), Leon Cooper (1930–   ), and John Schrieffer (1931–   ) – who invented it in 1957.

In 1937, the Russian physicist Peter Kapitza (1894–1984) discovered that when helium is cooled below 2.17 K (–270.98°C/–455.76°F), it becomes superfluid, flowing with zero resistance through tiny apertures, and flowing over the edges of any open container that it was kept in.

BCS was not adequate to explain the phenomenon of 'high-temperature' superconductivity, discovered in certain complex ceramic materials in 1986 by the German physicist Georg Bednorz (1950–　) and the Swiss physicist Karl Alex Müller (1927–　). The race was on to find materials that would superconduct at room temperatures, which could create enormous savings in electricity supply. Superconducting magnets have been utilized in many applications, perhaps most notably in body scanners used for medical diagnosis.

Particles, forces, and radiation

During the 20th century new forces were found to be at work in the heart of the atom. Radioactivity gave evidence of an astonishing new subatomic particle. When an electron or positron was emitted by a nucleus – a process called beta decay – some energy was generally found to be missing. In 1931 the Austrian-born Swiss physicist Wolfgang Pauli (1900–1958) suggested that this energy was being carried off by an unknown uncharged particle, which was later called the neutrino ('little neutral one'). Its mass must be extremely small and perhaps even zero. It reacted so rarely with other particles that one could travel through the entire Earth with only a tiny chance of interacting with any particle encountered on the way. In 1956 its existence was confirmed when floods of neutrinos from a nuclear reactor were seen to produce flashes of light in liquid. Trillions upon trillions of these particles flood through our bodies every second, completely unnoticed.

The interaction responsible for beta decay is called the weak nuclear force. It is 10 billion times weaker than electromagnetism, but 10^{30} times stronger than gravitation. But there had to be another, far stronger interaction at work in the nucleus. The positively charged protons in the atomic nucleus repel each other with an enormous force. Some other force, with a range limited to the dimensions of the nucleus, must overcome this repulsion.

In 1935 Japanese physicist Hideki Yukawa (1907–1981) suggested that protons and neutrons continually swapped identities by exchanging an electrically charged particle, now called the muon. Experimentalists thought they had found this particle when they discovered a new particle reaching the Earth's surface in cosmic rays from space. However, the muon, although 207 times as massive as the electron, was too light to be Yukawa's particle.

In 1947 the pi-meson, or pion, was discovered, with a mass 274 times that of the electron. This is the particle that is exchanged between protons and neutrons. The force that this exchange gives rise to is called the strong nuclear force, or simply the strong force. It brought to four the number of fundamental interactions.

Quantum electrodynamics

The idea of forces arising from 'messenger' particles being exchanged was used when quantum theory was applied to electromagnetism. The US physicists Richard Feynman (1918–1988) and Julian Schwinger (1918–1994) and the Japanese physicist Sin-Itiro Tomonaga (1906–1979) developed the theory of quantum electrodynamics (QED). QED deals with interactions between charged particles and between electromagnetic fields and charged particles. It treats electromagnetic interactions as the exchange of photons, which are the 'carriers' or 'messengers' of electromagnetic forces.

To understand the strong and weak forces, and the structure of the atomic nucleus, it was necessary to go deeper into the nucleus. Rutherford and Chadwick had bombarded atoms with naturally occurring alpha particles. The English physicists John Cockcroft (1897–1967) and Ernest Walton (1903–1995) improved on this by building the first particle accelerator, which used an accelerating potential of 800 kV to accelerate the protons and fire them at target atoms, producing in 1932 the first artificially induced nuclear reactions. In 1931 the US physicist Ernest O Lawrence (1901–1958) invented the cyclotron, in which charged particles were whirled in a spiral path, building up energy from a series of voltage kicks twice in each revolution. Such machines were developed into huge instruments. LEP, the Large Electron-Positron collider at CERN, the European Laboratory for Particle Physics near Geneva, Switzerland, was a ring 27 km/17 mi in circumference. Bunches of

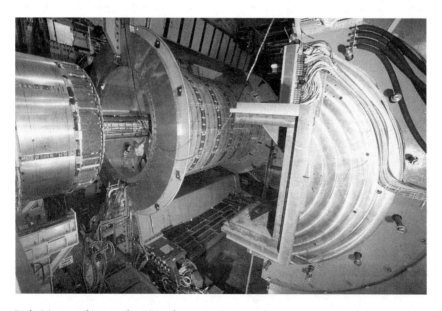

Technicians working on the OPAL detector, a large, multi-purpose particle detector, sited on the LEP accelerator at CERN, Switzerland, in 1993. Patrice Loiez/CERN

particles were accelerated to close to the speed of light, and then circulated, perhaps for hours, in storage rings, before being allowed to collide head-on with another stream of particles circulating in the opposite direction. LEP functioned for 11 years before being closed in November 2000.

The contributions of many theorists and large teams of experimentalists at the great accelerator laboratories built up a picture that is now called the standard model. According to the standard model, subatomic particles can be divided into three groups: hadrons, which interact through all the four fundamental interactions – gravitation, electromagnetism, and the weak and strong nuclear forces; leptons, which do not interact through the strong force, but do so through the three other interactions; and gauge bosons, or 'messenger' particles, which are 'carriers' of the interactions.

The hadrons are not fundamental particles but are made up of smaller particles called quarks, whose existence was suggested independently by the US physicists Murray Gell-Mann (1929–) and George Zweig in the early 1960s. We cannot see quarks singly because it would require far too much energy to tear one out of a hadron. However, their presence can be detected in the pattern of particles spraying out of a collision of particles. And it is possible that single quarks exist in cosmic rays, left over from the birth of the universe. A theory known as colour confinement suggests that the force between two quarks would be the same at a distance of 10^{-15} m as it is at 1 m. Current accelerators have the energy to separate quarks to a distance of the order of 10^{-13} m and it is estimated that separating two quarks to 1 m would require an increase in energy of the order of 10^{12} times.

Although the strong force between the protons and neutrons in a nucleus is 'carried' by pions, the pions themselves are made of pairs of quarks. The direct quark–quark interaction, in the nucleus and elsewhere, is carried by messenger particles called gluons.

Leptons are elementary particles as far as we know; that is, they do not seem to be made of any other particles. They fall into three 'generations': these comprise the electron, the muon, and a third particle, the tau; a neutrino associated with each of these – the electron–neutrino, the muon–neutrino, and the tau–neutrino; and the corresponding antiparticles of all these. In addition, each generation of leptons is matched with a pair of quarks. The three generations of leptons and quarks can be shown as:

Leptons	Quarks
electron neutrino	up
electron	down
muon neutrino	charm
muon	strange
tau neutrino	top
tau	bottom

As described above, electromagnetism is carried by photons, and the strong force is carried by gluons. The weak force is carried by messenger particles called the W and Z bosons. Physicists believe that gravitation is carried by particles that they have named gravitons, but they have not yet been detected.

During the 1960s the US physicists Sheldon Glashow (1932–) and Steven Weinberg (1933–), and the Pakistani physicist Abdus Salam (1926–1996) developed 'electroweak' theory, which gives a single description of the electromagnetic and weak interactions. Physicists speak of the electromagnetic and weak interactions as having been 'unified' by this work. Until his death, Einstein had tried and failed to weld gravitation and electromagnetism into a single force. Modern physicists now have four interactions to unify in this way. The electroweak theory is a step in this direction.

Energy from atoms

When radioactivity was discovered, physicists realized that there were vast amounts of energy locked away inside the atoms, far greater than is released in the rearrangements of electrons that occur in chemical reactions. There was talk of the possibility of harnessing this energy but no one knew how it could be done.

In 1948 the Russian-born US physicist George Gamow (1904–1968) pointed out that at very high temperatures and pressures the nuclei of light atoms could be 'fused', or melded together, to yield heavier nuclei, whose mass would be slightly less than the sum of the masses of the original nuclei. The mass difference would be converted into energy in accordance with relativity theory. Gamow proposed this as the process that powers the Sun and other stars. These are made up predominantly of hydrogen, the lightest atom. At the centre of the Sun, the temperature is 15 million°C, and the density is 160 times the density of water. Here hydrogen is converted to helium.

In 1938 a new source of nuclear energy was discovered. The German chemist Otto Hahn (1879–1968) and the Austrian physicist Lise Meitner (1878–1968) were puzzled by the strange results of an experiment in which they had bombarded uranium with neutrons. Meitner was forced by the takeover of her country by Nazi Germany to flee to Sweden. From there she published a paper, with her nephew Austrian-born British physicist Otto Frisch (1904–1979), suggesting a previously unsuspected type of nuclear breakdown, in which the nucleus of uranium (atomic number 92) broke into two nuclei of roughly equal size, such as barium (atomic number 56) and krypton (atomic number 36). Even before the paper appeared, her findings were reported informally in the USA by Niels Bohr while on a visit.

Several physicists realized that the neutrons released in fission could trigger other fissions in a chain reaction, releasing enormous amounts of energy. In 1939, Albert Einstein added his signature to a letter written by two other physicists, the Hungarian Leo Szilard (1898–1964) and Eugene Wigner, Hungarian-born but now a naturalized US citizen. The letter was addressed to US president Theodore Roosevelt, warning him that the fission of uranium presented the possibility of an explosive device of great power. President

Austrian-born Swedish physicist Lise Meitner and German physical chemist Otto Hahn in their laboratory in Germany in 1913. AIP Emilio Segrè Visual Archives

Roosevelt authorized a study of the question, in a research programme later called the Manhattan Project. In Britain, Otto Frisch and German-born British physicist Rudolf Peierls (1907–1995) alerted the government in early 1940. In late 1941, on the eve of the Japanese attack on Pearl Harbor, President Roosevelt authorized a major effort to discover such as device.

A team led by the Italian-born US physicist Enrico Fermi began constructing a nuclear 'reactor', which would release fission energy in a controlled manner. The Chicago reactor, built in a squash court beneath a sports stadium at the University of Chicago, USA, generated nuclear energy in December 1942. The building of bombs was carried out at Los Alamos, New Mexico, USA. The first 'atom bomb,' using plutonium, an artificial element of atomic number 94, was test-fired in the New Mexico desert on 16 July 1945. The second, using uranium, was dropped on the Japanese city of Hiroshima on 6 August 1945. Three days later the city of Nagasaki was destroyed by a plutonium bomb. Japan surrendered World War II on 14 August.

After the war many designs of reactor were built. They generated heat that was carried away by a coolant fluid – water, a liquid metal, or a gas – to generate steam to drive turbines. Some used U-235 as fuel, some used plutonium.

But hopes for the industry were blighted by growing public mistrust of nuclear energy. A series of accidents at nuclear stations in the USA, UK, and elsewhere eroded public confidence. On 26 April 1986, a nuclear reactor at Chernobyl, Ukraine, ran out of control and there was an explosion in the core. Officially 31 people died in the accident or from radiation poisoning later, but a radioactive cloud drifted across western Europe, leading to an unknown number of cancers in humans. Legal restrictions have led to the end of new reactor building in the USA, home of the world's major nuclear industry. In other countries, notably France, nuclear energy is still of major importance.

The 'hydrogen bomb', which exploits the other form of nuclear energy, fusion power, was first exploded on 1 November 1952. In the hydrogen bomb the temperature and pressure momentarily achieved by the explosion of a fission bomb causes nuclei of hydrogen to fuse to form helium. The first bomb was a few hundred times as powerful as the Hiroshima bomb. Later H-bombs were more powerful still.

The struggle to harness fusion energy for peaceful purposes still continues. Most designs are based on the Russian tokamak, a torus (hollow ring) containing a plasma of deuterium (heavy hydrogen, whose nucleus contains one proton and one neutron). This is contained by a magnetic field so that it does not touch the walls and lose heat. A US design uses crossed laser beams to create pulses of high temperature in a small region of plasma. Pulses of energy have been extracted in both these designs, holding out the promise of industrial-scale use in the future.

The 'mushroom cloud' of smoke rising 18,300 m/60,000 ft above Nagasaki after the dropping of the atom bomb on 8 August 1945. CORBIS

Physics and cosmology

While physicists were probing the atom, astronomers were extending their view across the universe. The US astronomer Edwin Hubble (1889–1953) calculated the distance to what was then called the Great Nebula in the galaxy Andromeda. He established that this spiral 'nebula' (a Latin word meaning 'mist') was not a cloud of gas within the Galaxy, our system of stars, but a galaxy in its own right. According to modern measurements, it is about 2 million light years from us. (A light year is the distance that light travels in one year; since light travels at 300,000,000 m per second, a light year is approximately 10^{16} miles. The reach of modern optical and radio telescopes has extended until now the farthest objects detectable are about 10 billion light years away.

Looking across this vast expanse of space is also looking back in time. We see the galaxy in Andromeda as it was 2 million years ago. Hubble analysed the light from the galaxies. He found that, apart from the closest galaxies, they all showed a red shift in their light. That is, dark and bright lines in the visible spectrum were shifted towards the red end of the spectrum. This meant that the wavelengths of the light were lengthened in comparison with the same light from a source on Earth. The wavelengths of the invisible radiations from the galaxies were also lengthened. The farther the galaxy was from us, as estimated by its apparent brightness, the greater its red shift.

Hubble interpreted the red shift as a result of the Doppler effect – the lengthening of a wave when the source is moving away from the observer. It is named after the 19th-century Austrian physicist Christian Johann Doppler (1803–1853). So the universe is expanding: the galaxies are rushing away from each other as if from some colossal primordial explosion. A Belgian churchman, the abbé Georges Lemaître (1894–1966), proposed that the universe had begun in the explosion of a single 'super-particle', which he called the 'primaeval atom'. The idea was developed by George Gamow. He suggested that the universe began with a hot plasma of protons, electrons, and neutrons. Collisions among the particles built up larger and larger nuclei by the same fusion process that Gamow had suggested is occurring in the stars today. However, according to modern theory, only the lightest elements were formed in this 'Big Bang': various forms of hydrogen, helium, and lithium. The heavier elements were built up billions of years later, in the stars.

In 1963 the first quasar was identified. These objects pour out energy equal to that of an entire galaxy of hundreds of billions of stars, yet this is generated in a volume about the size of our Solar System. The only explanation theorists could find for this amazing energy output was the black hole.

Quasars and black holes

A black hole is an object whose gravity is so intense that no matter or light can escape from it. A star whose mass is a few times greater than that of the Sun explodes at the end of its life and can leave a black hole remnant.

Much more massive black holes, lying at the centres of galaxies, power quasars. Stars and interstellar gas and dust fall into it and disappear from the view of the outside universe for ever. Inside the black hole, according to theory, occur some of the weirdest phenomena of Einstein's relativity, but any living creatures falling in would be torn apart in a fraction of a second before they could observe them. But matter on its way into the black hole forms a rotating disc of closely packed matter, glowing with frictionally generated heat. This intensely bright mass of matter is the quasar.

The physical theory that can take us in imagination to the brink of a black hole can also take us back to the birth of the universe. The Big Bang took place about 15 billion years ago. Cosmologists work out what happened at each succeeding instant by using the knowledge gained in particle accelerators, where the energies involved in particle collisions momentarily equal those prevailing at various stages of the Big Bang.

At the birth of the universe space, time, matter, and energy appeared together; the first split second was a chaos of radiation and particles existing fleetingly, in a universe swelling from the size of an atom. Though the universe had a limited volume, there was nothing outside it – space curved back on itself in a way that only the equations of relativity can describe. The cosmic fireball expanded, thinned, and cooled. Our theories can be applied only back to a time 10^{-43} seconds after the beginning: before this, time and space are ripped apart by the incredible density and temperature ($10^{-32°}$) of the universe. There was only one 'superforce' acting, and we have no theory to deal with it yet. After this time, gravity appeared as a separate force. At 10^{-35} seconds the strong force separated out. At 10^{-12} seconds electromagnetism split off from the weak force. The universe was a soup of exotic particles, including single quarks, as well as high-energy gamma photons.

When the universe was a millionth of a second old, its temperature had fallen to 10 trillion K, and quarks began to combine, forming protons, neutrons, and mesons. As the temperature fell still further these were able to combine and stay together, forming various types of light nuclei. This

Photograph of a black hole taken by NASA's Chandra X-ray Observatory. Gas from the companion star is drawn by gravity onto the black hole in a swirling pattern. As the gas nears the event horizon ('edge') of the black hole, a strong gravitational red shift makes it appear redder and dimmer. When the gas finally crosses the event horizon, it disappears from view. This is why the region within the event horizon appears black. CXC/M Weiss

process ceased after three minutes when the expanding plasma had thinned and cooled too much. It was a few hundred thousand years later that electrons fell into orbits around the nuclei to form atoms.

Physics in the 21st century

In the 21st century physics will become an ever costlier enterprise, demanding ever greater amounts of taxpayers' money. But the money will probably be forthcoming. Ever since nuclear energy was discovered and applied, governments have understood that fundamental physics research is a rich source of new developments that can enrich and strengthen nations. Physicists will require still more sophisticated and costly instruments. CERN's 27-km/17-mi ring housing LEP (which was shut down in 2000) will accommodate the LHC, the Large Hadron Collider, which will accelerate protons and other hadrons through a trillion volts. Beyond the orbit of the Moon the Next Generation Space Telescope (NGST) will probe the early universe, and its results are bound to pose new problems for physics. Satellite systems using an array of laser beams spanning 500 million m (5×10^8 m)/3 million mi may be set up in space to search for gravity waves – ripples in space-time from violent events in the universe, such as the explosion of stars or the merging of black holes.

With such equipment physicists will address the many problems that they were working on at the turn of the century. For example, are matter and antimatter true 'mirror-images' of one another? Experimental results suggesting that they behave slightly differently in certain interactions were reported in 1999. Such a difference could explain why antimatter had almost completely disappeared from the universe within fractions of a second of the Big Bang, despite the fact that matter and antimatter should have been produced in equal amounts initially. But physicists are sure that the matter we can directly observe is only a fraction of all the matter that exists. Some of the unknown 'dark matter' reveals its presence by its gravitational effects on the movements of galaxies. Some of it may be just cold, dark ordinary matter, mostly hydrogen gas. But physicists conjecture that it may consist of an undiscovered form of matter, WIMPs, or weakly interacting massive particles. These would interact with each other and with ordinary matter by gravitation alone. The search is on for such matter, both in the laboratory and in space.

Evidence was reported in 1998 that the expansion of the universe is accelerating. According to conventional physics the gravitational pull of the galaxies on each other should slow down the expansion. But if the new results are right, then there is a repulsive force between masses that increases with distance, and which is driving the galaxies apart ever faster. The 'new force' with 'repulsive' gravity is being called quintessence.

Even without such phenomena to complicate things, the standard model of elementary-particle interactions has many difficulties. For example, many quantities, such as the masses of the particles and the strengths of the interactions, are unexplained. The search is on for the Higgs boson, a particle whose existence was suggested by the English theoretical physicist Peter Higgs (1929–) and two Belgian physicists, Robert Brout and François Englert. If it is found it will explain how the masses of particles arise. The fact that elementary particles fall into three generations is another mystery that physicists hope to solve soon. Electromagnetism and the weak force have been 'unified' by electroweak theory; now it is necessary to unify this with the theory of the strong force, to form a grand unified theory, or GUT. Theorists are working intensively on string theory, according to which elementary particles are tiny vibrating 'strings', rather than points. According to string theory the universe has 10 or 11 dimensions, of which most 'rolled up' in the first instant of the Big Bang, leaving us with the three dimensions of space and one of time that we experience today. String theory offers the best hope of unifying all the fundamental interactions, including gravitation, to form the much heralded 'theory of everything'. But one thing is certain: even a 'theory of everything' will pose a vast range of new problems. The job of physicists will never be finished.

Chris Cooper

2 Chronology

1900

French physicist Antoine-Henri Becquerel demonstrates that beta particles are fast moving electrons.

French physiologist Paul Ulrich Villard discovers a new type of radiation that is later known as gamma rays.

German physicist Max Planck suggests that black bodies (perfect absorbers) radiate energy in packets or quanta, rather than continuously. He thus begins the science of quantum physics, which revolutionizes the understanding of atomic and subatomic processes.

German physicists Johann Phillip Elster and Hans F Geitel invent the first practical photoelectric cell.

1901

Becquerel reports the first radiation burn. It is caused by a sample of radium carried in his waistcoat pocket and leads to the use of radium for medical purposes.

1902

English physicist Oliver Heaviside and US electrical engineer Arthur Kennelly independently predict the existence of a conducting layer in the atmosphere that reflects radio waves.

Canadian-born US physicist Reginald Fessenden discovers the heterodyne principle whereby high-frequency radio signals are converted to lower frequency signals that are easier to control and amplify. It leads to the superheterodyne principle essential in modern radio and television.

Italian physicist Guglielmo Marconi discovers that radio waves are transmitted further at night than during the day because they are affected by changes in the atmosphere (actually by a layer of ionized gas in the ionosphere).

New Zealand-born British physicist Ernest Rutherford and English physical chemist Frederick Soddy discover thorium X and publish *The Cause and Nature of Radioactivity*, which outlines the theory that radioactivity involves the disintegration of atoms of one element to form atoms of another. Amongst other things, the discovery lays the foundation for radiometric dating of natural materials.

Polish-born French physicist Marie Curie and French chemist André-Louis Debierne isolate radium as a chloride compound.

Scottish physicist William Thomson (Lord Kelvin) proposes a model of the atom in which electrons are embedded in a sphere of positive charge.

1903

New Zealand-born British physicist Ernest Rutherford discovers that a beam of alpha particles is deflected by electric and magnetic fields. From the direction of deflection he is able to prove that they have a positive charge and from their velocity he determines the ratio of their charge to their mass. He also names the high-frequency electromagnetic radiation escaping from the nuclei of atoms as gamma rays.

Scottish chemist William Ramsay shows that helium is produced during the radioactive decay of radium – an important discovery for the understanding of nuclear reactions.

Scottish physicist Charles Thomson Rees Wilson develops the sensitive electroscope, which is used to detect the electric charge of ionizing radiation.

1904

English electrical engineer John Fleming patents the diode valve, which allows electricity to flow in only one direction. It is an essential development in the evolution of television.

German engineer Christian Hülsmeyer patents the first primitive radar system.

Japanese physicist Hantaro Nagaoka proposes a model of the atom in which the electrons are located in an outer ring and orbit the positive charge, which is located in a central nucleus. The model is ignored because it is thought the electrons would fall into the nucleus.

New Zealand-born British physicist Ernest Rutherford publishes *Radioactivity*, summarizing his work on the subject and pointing out that radioactivity produces more heat than chemical reactions do.

Marie Curie and her husband Pierre in their Paris laboratory. They received the Nobel Prize for Physics in 1903 for the discovery of radioactivity. In 1911 Marie Curie became the first person to be awarded the Nobel prize twice, when she was awarded the chemistry prize for her discovery of radium. Some of her notebooks are so radioactive that they cannot be handled. AEA Technology

1905

German-born US physicist Albert Einstein develops his special theory of relativity in a series of four papers in Switzerland. In 'On the Motion – Required by the Molecular Kinetic Theory of Heat – of Small Particles Suspended in a Stationary Liquid' he explains Brownian motion. In 'On a Heuristic Viewpoint Concerning the Production and Transformation of Light' he explains the photoelectric effect by proposing that light consists of photons and also exhibits wavelike properties. In 'On the Electrodynamics of Moving Bodies' he proposes that space and time are one and that time and motion are relative to the observer. In 'Does the Inertia of a Body Depend on its Energy Content?' he argues that mass and energy are equivalent, and can be expressed by the formula $E = mc^2$.

English chemist John William Strutt, Baron Rayleigh, invents the radium clock.

Scottish physicist William Thomson (Lord Kelvin) proposes a model of the atom in which positively and negatively charged spheres alternate.

1906

English physicist Charles Glover Barkla demonstrates that each element can be made to emit X-rays of a characteristic frequency.

German physicist Herman Nernst formulates the third law of thermodynamics, which states that matter tends towards random motion and that energy tends to dissipate at a temperature above absolute zero (–273.12°C/–350°F).

US physicist Lee De Forest invents the 'audion tube', a triode vacuum tube with a third electrode, shaped like a grid, between the cathode and anode that controls the flow of electrons and permits the amplification of sound. It is an essential element in the development of radio, radar, television, and computers.

1907

French physicist Pierre-Ernest Weiss develops the domain theory of ferromagnetism, which suggests that in a ferromagnetic material, such as lodestone, there are regions, or domains, where the molecules are all magnetized in the same direction. His theory leads to a greater understanding of rock magnetism.

1908

Dutch physicist Heike Kamerlingh Onnes liquefies helium at 4.2 K (–268.95°C/–452.11°F).

German physicist Hans Geiger and New Zealand-born British physicist Ernest Rutherford develop the Geiger counter, which counts individual alpha particles emitted by radioactive substances.

New Zealand-born British physicist Ernest Rutherford and his student Thomas Royds prove that the alpha particle is the helium nucleus.

US physicist Percy Williams Bridgman invents equipment that can create atmospheric pressures of 100,000 atmospheres (later 400,000), creating a new field of investigation.

1909

Dutch physicist Hendrik Lorentz publishes *The Theory of Electrons and its Applications to the Phenomena of Light and Radiant Heat,* in which he explains the production of electromagnetic radiation in terms of electrons.

German-born US physicist Albert Einstein introduces his idea that light exhibits both wave and particle characteristics.

1910

English physicist J J Thomson discovers the proton.

Polish-born French physicist Marie Curie publishes *Treatise on Radiography,* outlining her ideas about radioactivity.

German physicist Wolfgang Gaede develops the molecular vacuum pump, which can generate a vacuum of 0.00001 mm of mercury.

Swedish industrial engineer Nils Dalén designs an automatic sun valve to regulate a gaslight by the action of sunlight, turning it off at sunrise and on at sunset. It gains widespread use in uncrewed lighthouses and buoys.

French industrial chemist Georges Claude develops neon lighting when he discovers that the gas emits light when an electric current is passed through it. As it is initially only possible to produce red lighting, its potential is mainly restricted to advertising. However the techniques have been developed and a range of colours, in advertising signs, can be seen in any high street.

1911

Dutch physicist Heike Kamerlingh Onnes discovers superconductivity, the characteristic of a substance of displaying zero electrical resistance when cooled to just above absolute zero.

German-born US physicist Albert Einstein calculates the deflection of light caused by the Sun's gravitational field.

New Zealand-born British physicist Ernest Rutherford proposes the concept of the nuclear atom, in which the mass of the atom is concentrated in a nucleus occupying 1/10,000 of the diameter of the atom and which has a positive charge balanced by surrounding electrons.

New Zealand-born British physicist Ernest Rutherford and English chemist Frederick Soddy describe the 'transmutation' of the elements, in which a simpler atom is produced from a complex one.

US physicist Robert Millikan measures the electric charge on a single electron in his oil-drop experiment, in which the upward force of the electric charge on an oil droplet precisely counters the known downward gravitational force acting on it.

1912

German physicist Max von Laue demonstrates that crystals are composed of regular, repeated arrays of atoms by studying the patterns in which they diffract X-rays. It is the beginning of X-ray crystallography.

Scottish physicist Charles Thomson Rees Wilson perfects the cloud chamber, which detects ion trails since water molecules condense on ions. It is used to study radioactivity, X-rays, cosmic rays, and other nuclear phenomena.

1913

Danish physicist Niels Bohr proposes that electrons orbit the atomic nucleus in fixed orbits, thus upholding New Zealand-born British physicist Ernest Rutherford's model proposed in 1911.

English chemist Frederick Soddy coins the term 'isotope' (from the Greek *isos*, 'equal', and *topos*, 'place') to describe atoms of the same chemical element but with different atomic numbers.

English physicist J J Thomson develops a mass spectrometer, called a parabola spectrograph. A beam of charged ions is deflected by a magnetic field to produce parabolic curves on a photographic plate.

English physicist J J Thomson discovers neon-22, an isotope of neon. It is the first isotope of a nonradioactive element to be discovered.

English physicists William and Lawrence Bragg develop X-ray crystallography by establishing that the orderly arrangement of atoms in crystals displays interference and diffraction patterns. They also demonstrate the wave nature of X-rays.

German-born US physicist Albert Einstein formulates the law of photochemical equivalence, which states that for every quantum of radiation absorbed by a substance one molecule reacts.

German physicist Johannes Stark discovers that an electric field splits the spectral lines of hydrogen.

German physicists Hans Geiger and Ernst Marsden prove, by scattering experiments, that New Zealand-born British physicist Ernest Rutherford's model of the nuclear atom (proposed in 1911) is correct.

US physicist Albert Munsell develops the Munsell colour scheme, which defines colours according to their hue (wavelength), value (brightness), and chroma (purity).

1914

German physicists James Franck and Gustav Hertz provide the first experimental evidence for the existence of discrete energy states in atoms and thus verify Danish physicist Niels Bohr's atomic model.

US physicist Robert Millikan demonstrates that radiation shows some of the properties of particles – a verification of Einstein's photoelectric effect.

Work of the National Physical Laboratory, UK, is extended to include the testing and certification of radium preparations.

1916

German-born US physicist Albert Einstein publishes *The Foundation of the General Theory of Relativity*, in which he postulates that space is curved locally by the presence of mass and that this can be demonstrated by observing the deflection of starlight around the Sun during a total eclipse. This replaces previous Newtonian ideas which invoke a force of gravity.

US physical chemist William David Coolidge patents an X-ray tube that can produce highly predictable amounts of radiation. It serves as the prototype of the modern X-ray tube.

1918

English astrophysicist Arthur Eddington publishes *Report on the Relativity Theory of Gravitation*, which is the first explanation of German-born US physicist Albert Einstein's theory of relativity in English.

1919

Aston confirms that the element neon consists of two isotopes of different masses. Their relative abundances explain the observed atomic weight of 20.25.

New Zealand-born British physicist Ernest Rutherford splits the atom by bombarding a nitrogen nucleus with alpha particles, discovering that it ejects hydrogen nuclei (protons). It is the first artificial disintegration of an element and inaugurates the development of nuclear energy.

29 May

English astrophysicist Arthur Eddington and others observe the total eclipse of the Sun on Príncipe Island (West Africa), and discover that the Sun's gravity bends the light from the stars beyond the edge of the eclipsed sun, thus confirming German-born US physicist Albert Einstein's theory of relativity.

1920

Danish physicist Niels Bohr introduces the correspondence principle, which relates the motion of particles to the radiation emitted.

English chemist Frederick Soddy suggests that isotopes can be used to determine geological age.

New Zealand-born British physicist Ernest Rutherford recognizes the hydrogen nucleus as a fundamental particle and names it the 'proton'.

1921

English physicist Patrick Stuart Blackett uses cloud-chamber photographs of atomic nuclei bombarded with alpha particles to show how they are disintegrated.

The total eclipse of the Sun on 29 May 1919 that verified one of the predictions contained in Einstein's general theory of relativity, that rays of light are affected by gravitation. Royal Astronomical Society

German physicist Max Born develops a mathematical description of the first law of thermodynamics.

1922

English physicist Patrick Maynard Blackett undertakes experiments on the transmutation of elements.

US physicist Arthur Holly Compton discovers that X-rays scattered by an atom have a shift in frequency. He explains the phenomenon, known as the Compton effect, by treating the X-rays as a stream of particles, thus confirming the wave–particle idea of light.

1923

German-born US physicist Albert Michelson measures the speed of light, obtaining a value close to the modern value of 299,792 km per second 186,282 mi per second.

1924

English physicist Edward Appleton discovers that radio emissions are reflected by an ionized layer of the atmosphere.

French physicist Louis de Broglie argues that particles can also behave as waves, laying the foundations for wave mechanics. He demonstrates that a beam of electrons has a wave motion with a short wavelength. The discovery permits the development of the electron microscope.

1926

Austrian physicist Erwin Schrödinger develops wave mechanics.

Dutch physicist Willem Keesom solidifies helium.

German-born US physicist Albert Michelson uses an eight-sided rotating prism to determine the speed of light. The value determined by Michelson is 2.99774×10^8 m per second.

1927

German physicist Werner Heisenberg propounds the 'uncertainty principle' in quantum physics, which states that it is impossible to simultaneously determine the position and momentum of an atom. It explains why Newtonian mechanics is inapplicable at the atomic level.

German physicists Walter Heitler and Fritz London discover that quantum mechanics can explain chemical bonding.

US physicist Clinton Davisson and, independently, English physicist George Paget Thomson show that electrons can be diffracted.

1928

English physicist Paul Dirac describes the electron by four wave equations. The equations imply that the electron must spin on its axis and that negative states of matter must exist.

German physicist Rolf Wideröe develops the resonance linear accelerator, which he uses to accelerate potassium and sodium ions to an energy of 710 keV to split the lithium atom.

Russian physicist George Gamow shows that the atom can be split using low-energy ions. It stimulates the development of particle accelerators. Gamow and US physicists Ronald W Gurney and Edward Condon explain the relationship between the half-life of a radioactive element and the energy emitted by the alpha particle.

Einstein publishes *Unitary Field Theory*, in which he attempts to explain the various atomic forces by a single theory.

Irish physicist Ernest Walton and English physicist John D Cockcroft develop the first particle accelerator.

Russian physicist George Gamow, US physicist Robert Atkinson, and German physicist Fritz Houtermans suggest that thermonuclear processes are the source of solar energy.

1931

US chemist Harold C Urey and atomic physicist J Washburn discover that electrolysed water is denser than ordinary water, leading to the discovery of deuterium ('heavy hydrogen'). The discovery ushers in the modern field of stable isotope geochemistry.

US physicists Ernest Lawrence and M Stanley Livingston build a cyclotron, particle accelerator.

US physicist Ernest Lawrence beside the first cyclotron to be too large to be laid upon a lecture table. University of California Lawrence Radiation Laboratory, AIP Emilio Segrè Visual Archives

1932

British physicist James Chadwick discovers the neutron, an important discovery in the development of nuclear reactors.

British physicists John D Cockcroft and Ernest Walton develop a high-voltage particle accelerator, which they use to split lithium atoms.

The shape-memory effect is discovered by US chemical engineers L C Change and T A Read in a gold and cadmium alloy; if the alloy is bent it returns to its original shape when heated.

1933

German physicists Walter Meissner and R Ochensfeld discover that super-conducting materials expel their magnetic fields when cooled to superconducting temperatures – the Meissner effect.

US physicist Carl D Anderson succeeds in producing positrons by gamma irradiation.

c. 1934

Italian physicist Enrico Fermi bombards uranium with neutrons and discovers the phenomenon of atomic fission, the basic principle of atomic bombs and nuclear power.

1934

Italian physicist Enrico Fermi suggests that neutrons and protons are the same fundamental particles in two different quantum states.

Russian physicist Pavel Cherenkov discovers that light is emitted when particles pass through liquids or transparent solids faster than the speed of light in the same medium. The phenomenon becomes known as 'Cherenkov radiation'.

1935

Canadian physical chemist William Giauque cools liquid helium to 0.0004°C/0.0002°F above absolute zero.

Japanese physicist Hideki Yukawa proposes the existence of a new particle, the meson, to explain nuclear forces.

1936

US physicist Carl D Anderson discovers the muon, an electron-like particle over 200 times more massive than an electron.

1938

Russian physicist Peter Kapitza discovers that liquid helium exhibits super-fluidity, the ability to flow over its containment vessel without friction, when cooled below 2.18K/−270.97°C.

German physicists Hans Bethe and Carl von Weizsäcker propose that nuclear fusion of hydrogen is the source of a star's energy.

French physicists Frédéric Joliot and Irène Curie-Joliot demonstrate the possibility of a chain reaction when they split uranium nuclei.

1940

Italian-born US physicist Emilio Gino Segrè and US physicists Dale Corson and K R Mackenzie synthesize astatine (atomic number 85).

The Rockefeller Foundation grants funds to the University of California, USA, to build a giant cyclotron (a type of particle accelerator), under the direction of US physicist Ernst Orlando Lawrence.

US physicist John R Dunning leads a research team that uses a gaseous diffusion technique to isolate uranium-235 from uranium-238. Because uranium-235 readily undergoes fission into two atoms, and in doing so releases large amounts of energy, it is used for fuelling nuclear reactors.

US physicists Edwin McMillan and Philip Abelson synthesize the first transuranic element, neptunium (atomic number 93), by bombarding uranium with neutrons at the cyclotron at Berkeley, California, USA.

1941

US physicists Glenn Seaborg and Edwin McMillan synthesize plutonium (atomic number 94).

1945

US physicist Edwin McMillan and Soviet physicist Vladimir Veksler (1943) independently describe the principle of phase stability. By removing an apparent limitation on the energy of particle accelerators for protons, it makes possible the construction of magnetic-resonance accelerators, or synchrotrons. Synchrocyclotrons are soon built at the University of California, USA, and in England.

1946

Russian physicist Igor Kurchatov creates the first Soviet atomic chain reaction.

English physicist Edward Appleton discovers the 'Appleton' layer in the ionosphere, one of the layers that reflect radio waves and make long-range radio communication possible.

1947

English physicist Patrick Maynard Blackett advances the theory that all massive rotating bodies are magnetic.

British physicists Cecil Frank Powell and Giuseppe Occhialini discover the pion (or pi-meson) subatomic particle.

US chemist and physicist Willard Libby develops carbon-14 dating.

1948

Hungarian-born British physicist Dennis Gabor invents holography, the production of three-dimensional images.

US physicists John Bardeen, William Bradley Shockley, and Walter Brattain develop the transistor in research at Bell Telephone Laboratories in the USA. A solid-state mechanism for generating, amplifying, and controlling electrical impulses, it revolutionizes the electronics industry by enabling the miniaturization of computers, radios, and televisions, as well as the development of guided missiles.

Russian-born US physicist George Gamow and US physicist Ralph Alpher develop the 'Big Bang' theory of the origin of the universe, which says that a primeval explosion led to the universe expanding rapidly from a highly compressed original state.

1950

British physicist Louis Essen develops a new means of calculating the velocity of light by measuring the speed of radio waves in a vacuum.

The US Atomic Energy Commission separates plutonium from pitchblende concentrates.

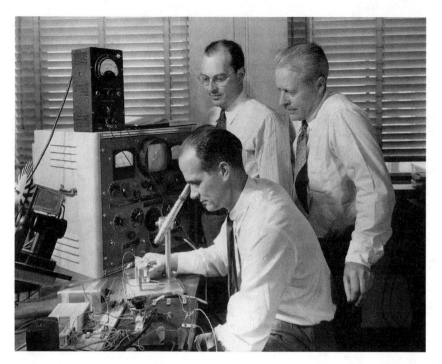

From left to right, US physicists William Shockley, John Bardeen, and Walter Brattain working together in 1948. They shared the Nobel Prize for Physics in 1956 for their work on the development of the transistor. Bell Laboratories, courtesy AIP Emilio Segrè Visual Archives

1951

US physicist Edward Purcell discovers line radiation (radiation emitted at only one specific wavelength) of wavelength 21 cm/8.3 in emitted by hydrogen in space. The discovery allows the distribution of hydrogen clouds in galaxies and the speed of the Milky Way's rotation to be determined.

1952

US nuclear physicist Donald Glaser develops the bubble chamber to observe the behaviour of subatomic particles. It uses a superheated liquid instead of a vapour to track particles. *See illustration on page 40.*

1953

An international laboratory for nuclear research is opened at Meyrin, near Geneva, Switzerland. It becomes known as CERN.

The screen of a bubble chamber detector at Fermilab, Illinois, showing the trails left by subatomic particles. Kevin Fleming/CORBIS

US physicist Murray Gell-Mann introduces the concept of 'strangeness', a property of subatomic particles, to explain their behaviour.

US physicist Robert Hofstadter discovers that protons and neutrons have an internal structure.

1956

US physicists B Cook, G R Lambertson, O Piconi, and W A Wentzel discover the antineutron by passing an antiproton beam through matter.

US physicists Clyde Cowan and Fred Reines detect the existence of the neutrino, a particle with no electric charge and no mass, at the Los Alamos Laboratory, New Mexico, USA.

1957

Japanese physicist Leo Esaki discovers tunnelling, the ability of electrons to penetrate barriers by acting as waves.

US physicists John Bardeen, Leon Cooper, and John Schrieffer formulate a theory to explain superconductivity, the characteristic of a solid material of losing its resistance to electric current when cooled below a certain temperature.

1958

US physicists Richard Feynman and Murray Gell-Mann formulate a theory that explains the phenomena associated with the weak interaction of sub-atomic particles.

1959

US physicist Luis Alvarez discovers the neutral xi-particle.

29 December

US theoretical physicist Richard Feynman delivers a paper entitled 'There's Plenty of Room at the Bottom' to the American Physical Society, in which he describes the manufacture of transistors and other electronic components one atom at a time. It anticipates nanotechnology.

1961

US physicist Murray Gell-Mann and Israeli physicist Yuval Ne'eman independently propose a classification scheme for subatomic particles that comes to be known as the Eightfold Way.

1962

Welsh physicist Brian Josephson discovers the Josephson effect, actually several effects, including the high-frequency oscillation of a current between two superconductors across an insulating layer. Experimental computers built in the 1980s using the Josephson junction are 10–100 times faster than conventional ones.

1964

Scientists at the Brookhaven National Laboratory, Upton, Long Island, USA, discover the fundamental omega-minus particle using the 'Nimrod' cyclotron.

US physicists James Cronin and Val Fitch discover an asymmetry in the behaviour of elementary particles, which is termed CP (charge parity) nonconservation.

US physicists Murray Gell-Mann and George Zweig independently suggest the existence of the quark, the building block of hadrons, which are subatomic particles that experience the strong nuclear force.

1967

A tank containing 100,000 gallons of cleaning fluid is installed in a former gold mine in South Dakota, USA, to detect neutrinos (uncharged elementary particles) from the Sun.

US nuclear physicists Sheldon Lee Glashow and Steven Weinberg and Pakistani nuclear physicist Abdus Salam separately develop the electroweak unification theory, which provides a single explanation for electromagnetic and 'weak' nuclear interactions.

March

The Fermi National Accelerator Laboratory ('Fermilab') is established near Chicago, Illinois, USA.

1970

US physicist Sheldon Glashow and associates postulate the existence of a fourth quark, which they name charm.

1971

English theoretical physicist Stephen Hawking suggests that after the Big Bang, mini black holes no bigger than protons but containing more than a billion tonnes of mass were formed and that they were governed by both the laws of relativity and of quantum mechanics.

1971–85

The optical parametric oscillator is developed by numerous physicists over a 15-year period; it uses photons to measure small movements of particles due to gravitational waves.

US physicist Murray Gell-Mann presents the theory of quantum chromodynamics (QCD), which explains how quarks interact. Strongly interacting particles consist of quarks, which are bound together by gluons.

1973

Researchers at the European Laboratory for Particle Physics (CERN) in Switzerland find some confirmation for the electroweak force – one of the fundamental forces – in neutrino reactions.

1974

US physicist Kenneth Geddes Wilson develops a technique for improving theories concerning transformations of matter called phase transitions.

US theoretical physicist Sheldon Glashow proposes the first grand unified theory – one that envisages the strong, weak, and electromagnetic forces as variants of a single superforce.

June

Soviet physicist Georgy Flerov and his associates announce the discovery of seaborgium (atomic number 106).

16 November

US physicists Burton Richter and Samuel Chao Chung Ting announce they have separately discovered the J/ψ particle, which consists of a charm quark combined with its antiparticle. It confirms the existence of the charm quark.

1975

'U' sub-atomic particles are produced at the accelerator at Stanford University, California, USA.

US physicist Samuel Hurst's machine identifies a single caesium atom in a sample of 10^{19} argon atoms.

August

US physicist Martin Perl discovers the tau lepton (or tauon), a fundamental particle.

1977

German-born US physicist Leon Lederman discovers the fifth, 'bottom', quark, combined with its antiquark in a particle called the upsilon meson.

1979

Physicists in Hamburg, West Germany, at DESY (Deutsches Elektron Synchroton) observe gluons – particles that carry the strong nuclear force that holds quarks together.

US physicist Arthur Ruoff subjects the gas xenon to a pressure of 320,000 atmospheres and obtains metallic xenon; it opens the way to the production of metallic hydrogen used in superconductivity research.

1980

German physicist Hans Dehmelt and colleagues at the University of Heidelberg, Germany, succeed in photographing a single barium atom.

1982

29 August
Scientists at Darmstadt, West Germany, announce the production of the element unnilenium (atomic number 109, now called meitnerium) by fusing bismuth and iron nuclei.

1983

The first experiments are conducted with the Joint European Torus (JET) equipment at Culham, UK, in an attempt to generate electricity by means of nuclear fusion.

June
The W and Z subatomic particles are detected in experiments at the European Laboratory for Particle Physics (CERN), Switzerland, by Italian physicist Carlo Rubbia and Dutch physicist Somin van der Meer; the existence of these particles had been predicted as carriers of the weak nuclear force.

1984

A team of international physicists at the European Laboratory for Particle Physics (CERN) in Geneva, Switzerland, discovers the sixth (top) quark; its discovery completes the theoretical scheme of subatomic building blocks.

An international team of scientists at Deutsche Elektronen-Synchrotron (DESY) near Hamburg, West Germany, discovers the 'zeta' subatomic particle, a massive particle with neutral charge.

German physicist Peter Armbruster and colleagues at the Heavy Ion Research Institute (GSI), Darmstadt, Germany, synthesize the element hassium (atomic number 108).

1986

IBM researchers develop the first scanning tunnelling electron microscope. It is built on a single silicon chip only 200 μm across.

Scientists use 10 laser beams, which deliver a total power of 100 trillion watts during one-billionth of a second, to convert a small part of the hydrogen nuclei contained in a glass sphere to helium at the Lawrence Livermore National Laboratory in California, USA; it is the first fusion reaction induced by a laser.

1989

March

US physicist Stanley Pons and English physicist Martin Fleischmann announce that they have achieved nuclear fusion at room temperature (cold fusion); other scientists fail to replicate their results.

1990

Researchers at IBM's Almaden Research Center in California, USA, are the first to manipulate individual atoms on a surface; they use a scanning tunnelling microscope and spell out the initials 'IBM'.

1 January

The volt, which measures electrical potential, and the ohm, which measures electrical resistance, are redefined in atomic terms.

1992

Two Japanese researchers develop a material which becomes superconducting at the relatively high temperature of $-103°C/-153°F$.

1993

Physicists at Lancaster University, UK, achieve a temperature of 2.8×10^{-10} K (0.28 billionths of a degree above absolute zero).

US physicists achieve the most successful nuclear fusion experiment to date when hydrogen isotopes are heated to 300 million degrees, creating 5.6 million watts of power.

1994

September

The European Synchrotron Radiation Facility (ESRF) opens in Grenoble, France.

1995

Australian physicists incorporate the world's largest niobium ingot (1.5 tonnes in weight and 3 m/9.8 ft long) into a gravity wave detector.

Physicists at the CERN research centre in Switzerland detect 'glueballs', gluons that stick together and behave as single particles.

The Los Alamos National Laboratory, New Mexico, USA, produces a flexible superconducting film that is superconducting at the relatively high temperature of 77 K ($-196°C/-320°F$).

US scientists at Fermilab, near Chicago, Illinois, USA, announce the discovery of the top quark, an elementary particle almost as heavy as a gold atom.

June

US physicists announce the discovery of a new form of matter, called a Bose–Einstein condensate (because its existence had been predicted by German-born US physicist Albert Einstein and Indian physicist Satyendra Bose), created by cooling rubidium atoms to just above absolute zero.

December

Scientists at the Heavy Ion Research Institute (GSI), Darmstadt, Germany, synthesize unununium (atomic number 111).

1996

Construction of the world's largest neutrino detector, the Antarctic Muon and Neutrino Detector Array (AMANDA), begins at the South Pole.

4 January

A team of European physicists at the CERN research centre in Switzerland create the first atoms of antimatter: nine atoms of antihydrogen survive for 40 nanoseconds.

February

Scientists at the Heavy Ion Research Institute (GSI), Darmstadt, Germany, synthesize ununbium (atomic number 112). A single atom is created, which lasts for a third of a millisecond.

1997

11 September

The Israeli physicist Rafi de-Picciotto demonstrates the formation, within semiconductor materials, of 'quasiparticles', which have a charge one-third that of an electron. They challenge the idea that charge always comes in discrete units based on the charge of a single electron.

1998

US astronomers Saul Perlmutter and Alex Filippenko suggest that not only is the universe expanding but that the rate of expansion is increasing.

11 July

Researchers at the Fermi National Accelerator Laboratory, Illinois, USA, announce the discovery of the tau neutrino, the least stable elementary particle of the lepton class.

1999

February

Scientists succeed in slowing down the speed of light from its normal speed of 299,792 km/186,282 mi per second to 61 km/38 mi per hour, opening up potential for the development of high-precision computer and telecommunications technologies, as well as for the advanced study of quantum mechanics.

2000

Physicists and astronomers start the search for 'quintessence'. Quintessence has the property of being able to cause the rate of expansion of the universe to increase. It is believed to achieve this feat by having 'repulsive gravitational force'.

3   Biographical Sketches

Alfvén, Hannes Olof Gösta (1908–1995)

Swedish astrophysicist. He shared the Nobel Prize for Physics in 1970 for his fundamental contributions to plasma physics, particularly in the field of magnetohydrodynamics (MHD) – the study of plasmas in magnetic fields.

Alfvén was born in Norrköping, Sweden, and educated at the University of Uppsala. In 1940 he joined the Royal Institute of Technology, Stockholm, later dividing his academic career between that and the University of California, San Diego, USA, from which he obtained a professorship in 1967. During the 1960s, he became involved in the Pugwash movement (founded by Bertrand Russell among others) warning of the dangers of nuclear proliferation. In 1972 he was among a group of Oxford scientists who appealed to governments to abandon fast-breeder nuclear reactors and concentrate efforts on nuclear fusion.

Alfvén formulated the frozen-in-flux theorem, according to which a plasma is – under certain conditions – bound to the magnetic lines of flux passing through it; later he used this theorem to explain the origin of cosmic rays. In 1939 he proposed a theory to explain aurorae and magnetic storms, which greatly influenced later ideas about the Earth's magnetosphere.

In 1942 Alfvén postulated that a form of electromagnetic wave would propagate through plasma; other scientists later observed this phenomenon in plasmas and liquid metals. Also in 1942 he developed a theory that the planets in the Solar System were formed from material captured by the Sun from an interstellar cloud of gas and dust.

Alvarez, Luis Walter (1911–1988)

US physicist. He led the research team that discovered the Ξ_0 subatomic particle in 1959. He also made many other breakthroughs in fundamental physics, accelerators, and radar. He worked on the US atom bomb for two years, at Chicago, Illinois, and Los Alamos, New Mexico, during World War II. He was awarded the Nobel Prize for Physics in 1968 for his work in elementary-particle physics, and discovery of resonance states, using the hydrogen bubble chamber and data analysis.

In 1980 Alvarez was responsible for the theory that dinosaurs disappeared because a meteorite crashed into Earth 65 million years ago, producing a dust cloud that blocked out the Sun for several years, causing dinosaurs and plants to die. The first half of the hypothesis is now widely accepted.

Alvarez was born in San Francisco and studied at Chicago. In 1945 he became professor at the University of California, working at the Lawrence Livermore Radiation Laboratory there 1954–59. During World War II he moved to the Massachusetts Institute of Technology, where he developed the VIXEN radar for the airborne detection of submarines, phased-array radars, and ground-controlled approach radar that enabled aircraft to land in conditions of poor visibility. He also participated in creating the atomic bomb dropped on Hiroshima, Japan.

Alvarez built the first practical linear accelerator and an accelerator for breeding plutonium, and invented the tandem electrostatic accelerator. He also devised, but never built, the microtron for accelerating electrons. In 1953 Alvarez met US nuclear physicist Donald Glaser, inventor of the bubble-chamber detector for subatomic particles. Alvarez decided to build a much larger chamber than Glaser had used, and to fill it with liquid hydrogen. He also developed automatic scanning and measuring equipment whose output could be stored on punched cards and then analysed using computers. Alvarez and co-workers used the bubble chamber to discover a large number of new short-lived particles. These experimental findings were crucial in the development of the 'eightfold way' model of elementary particles and, subsequently, the theory of quarks.

Anderson, Carl David (1905–1991)

US physicist. He shared the Nobel prize for Physics in 1936 for his discovery in 1932 of the positive electron (positron). His discovery of another particle, the muon, in 1937 launched elementary-particle physics.

Anderson was born in New York and educated at the California Institute of Technology, where he spent his entire career.

Using a modified cloud chamber of his own devising, Anderson found that positive electrons, or positrons, were present in cosmic rays, energetic particles reaching Earth from outer space. For this discovery, Anderson was awarded the Nobel prize. In 1937, Anderson found a new particle in cosmic rays, one with a mass between that of an electron and a proton. The new particle was first called a mesotron and then a meson muon. The muon was the first elementary particle to be discovered beyond the constituents of ordinary matter (proton, neutron, and electron).

US physicist Carl Anderson in 1935 next to the control panel of his cloud chamber, from the trailer at Pike's Peak, Colorado, where Anderson drove it in search of high altitudes.
AIP Emilion Segrè Visual Archives

Appleton, Edward Victor (1892–1965)

British physicist. He worked at Cambridge under Ernest Rutherford from 1920. He proved the existence of the Kennelly–Heaviside layer (now called the E layer) in the atmosphere, and the Appleton layer beyond it, and was involved in the initial work on the atom bomb. He was awarded the Nobel Prize for Physics in 1947 for his work on the physics of the upper atmosphere. He was made KCB in 1941 and GBE in 1946.

Appleton was born in Bradford, West Yorkshire, and educated at Cambridge. He became interested in radio as signals officer during World War I, and his research into the atmosphere was of fundamental importance to the development of radio communications. He was professor at King's College, London 1924–36, and at Cambridge 1936–39. He was secretary of the Department of Scientific and Industrial Research 1939–49, and principal and vice chancellor of Edinburgh University 1949–65.

By periodically varying the frequency of the British Broadcasting Corporation (BBC) transmitter at Bournemouth and measuring the intensity of the received transmission 100 km/62 mi away, Appleton found that there was a regular fading in and fading out of the signals at night but that this effect diminished considerably at dawn as the Kennelly–Heaviside layer broke up. Radio waves continued to be reflected by the atmo-sphere during the day but by a higher-level ionized layer. By 1926 this layer, which Appleton measured at about 230 km/145 mi above the Earth's surface (the first distance measurement made by means of radio), became generally known as the Appleton layer (it is now also known as the F layer).

Aston, Francis William (1877–1945)

English physicist who developed the mass spectrometer, which separates isotopes by projecting their ions (charged atoms) through a magnetic field. He was awarded the Nobel Prize for Chemistry in 1922 for his contribution to analytical chemistry and the study of atomic theory .

Aston was born and educated in Birmingham. From 1910 he worked in the Cavendish Laboratory, Cambridge, where J J Thomson was investigating positive rays from gaseous discharge tubes. Thomson and Aston examined the effects of electric and magnetic fields on positive rays, showing that the rays were deflected depending on their mass. The deflected rays were made to reveal their positions by aiming them at a photographic plate. The image produced on the photographic plate became known as a mass spectrum, and the instrument itself as a mass spectrometer. This became an essential tool in the study of nuclear physics and later found application in the determination of the structures of organic compounds.

Aston first examined neon gas and found that it consists of two isotopes. Over the next few years he examined the isotopic composition of more than 50 elements, and published *Isotopes* in 1922.

Bardeen, John (1908–1991)

US physicist. He was awarded the Nobel Prize for Physics in 1956 (with US physicists Walter Brattain and William Shockley) for the development of the transistor in 1948 and he became the first double winner of the Nobel Prize for Physics in 1972 (with Leon Cooper and Robert Schrieffer) for his work on superconductivity.

Bardeen was born in Madison, Wisconsin, and educated at the University of Wisconsin.

At the Bell Telephone Laboratory, New Jersey, 1945–51, in a team with Shockley and Brattain, Bardeen studied semiconductors, especially germanium, used in radar receivers in the same way that crystals had been used in the earliest radio sets. The work led to the development of the transistor in 1956. The second Nobel Prize was won for explaining superconductivity, the total loss of electrical resistance by some metals when cooled within a few degrees of absolute zero. The theory developed in 1957 by Bardeen, Schrieffer, and Cooper states that superconductivity arises when electrons travelling through a metal interact with the vibrating atoms of the metal.

Barkla, Charles Glover (1877–1944)

English physicist who studied the phenomenon of secondary radiation (the effect whereby a substance subjected to X-rays re-emits secondary X-radiation) and X-ray scattering. He was awarded the Nobel Prize for Physics in 1917 for his discovery that X-ray emissions are a form of transverse electromagnetic radiation, like visible light, and monochromatic.

Barkla was born in Widnes, Lancashire, and studied in Liverpool and at Cambridge. He was professor of physics at King's College, London, 1909–13, when he became professor of natural philosophy at Edinburgh University.

In 1903 Barkla published his first paper on secondary radiation. He found that the more massive an atom, the more charged particles it contains, and it is these charged particles that are responsible for the X-ray scattering. Barkla was one of the first to emphasize the importance of the amount of charge in an atom (rather than merely its atomic mass) in determining an element's position in the periodic table.

Between 1904 and 1907 Barkla found that, unlike the low atomic mass elements, the heavy elements produced secondary radiation of a longer wavelength than that of the primary X-ray beam, and that the radiation from the heavier elements is of two characteristic types. Barkla named the two types of characteristic emissions the K-series (for the more penetrating emissions) and the L-series (for the less penetrating emissions). He later predicted that an M-series and a J-series of emissions with different penetrances might exist, and an M-series was subsequently discovered.

Bednorz, Johannes Georg (1950–)

German physicist who, with Alexander Müller, was awarded the Nobel Prize for Physics in 1987 for his discovery of high-temperature superconductivity in ceramic materials. The discovery of these materials contributed towards the use of superconductors in computers, magnetic levitation trains, and the more efficient generation and distribution of electricity.

Bednorz and Müller showed in 1986 that a ceramic oxide of lanthanum, barium, and copper became superconducting at temperatures above 30 K, much warmer than for any previously known superconductor.

Bednorz was born in North-Rhine Westphalia, in the Federal Republic of Germany. In 1977 he started his doctoral studies at the Swiss Federal Institute of Technology in Zürich. He joined the IBM Zürich Research Laboratories at Rüschlikon in 1982.

Bell, John (1928–1990)

Northern Irish physicist who in 1964 discovered a paradoxical aspect of quantum theory: two particles that were once connected are always afterwards interconnected even if they become widely separated. As well as investigating fundamental problems in theoretical physics, Bell contributed to the design of particle accelerators.

Bell worked for 30 years at CERN, the European research laboratory near Geneva, Switzerland. He put forward mathematical criteria that had to be obeyed if the connection required by quantum theory really existed. In the early 1980s, a French team tested Bell's criteria, and a connection between widely separated particles was detected.

Bell Burnell, (Susan) Jocelyn (1943–)

Northern Irish astronomer. In 1967 she discovered the first pulsar (rapidly flashing star) with British radio astronomer Antony Hewish and colleagues at the Mullard Radio Astronomy Observatory, Cambridge, UK.

Jocelyn Bell was born in Belfast, near the Armagh Observatory, where she spent much time as a child. She was educated at Glasgow and Cambridge universities in Scotland and England. It was while a research student at Cambridge that the discovery of pulsars was made. Between 1968 and 1982 she did research in gamma-ray astronomy at the University of Southampton and in X-ray astronomy at the Mullard Space Science Laboratory, University College London. Then she worked on infrared and optical astronomy at the Royal Observatory, Edinburgh, Scotland. In 1991 she was appointed professor of physics at the Open University, Milton Keynes.

Bell spent her first two years in Cambridge building a radio telescope that was specially designed to track quasars. The telescope had the ability to record rapid variations in signals. In 1967 she noticed an unusual signal, which turned out to be composed of a rapid set of pulses that occurred precisely every 1.337 seconds. One attempted explanation of this curious phenomenon was that it emanated from an interstellar beacon, so initially it was nicknamed LGM, for Little Green Men. Within a few months, however, Bell located three other similar sources. They too pulsed at an extremely regular rate but their periods varied over a few fractions of a second and they all originated from widely spaced locations in our galaxy. Thus it seemed that a more likely explanation for the signals was that they were being emitted by a special kind of star – a pulsar.

Blackett, Patrick Maynard Stuart (1897–1974)

English physicist. He was awarded the Nobel Prize for Physics in 1948 for work in cosmic radiation and his perfection of the Wilson cloud chamber, an apparatus for tracking ionized particles, with which he confirmed the existence of positrons.

Blackett was born in Croydon, Surrey, and joined the navy in 1912; after World War I, he studied science at Cambridge. He held posts at various British academic institutions.

In 1924, working under physicist Ernest Rutherford at Cambridge, Blackett made the first photograph of an atomic transmutation, which was of nitrogen into an oxygen isotope. He continued to develop the cloud chamber and in 1932 designed one where photographs of cosmic rays were taken automatically. Later he discovered particles with a lifespan of 10^{-10} seconds, which became known as strange particles.

In the 1950s he turned to the study of rock magnetism. He was made a life peer, becoming Baron Blackett, in 1969.

Bohr, Niels Henrik David (1885–1962)

Danish physicist who was awarded the Nobel Prize for Physics in 1922 for his discovery of the structure of atoms and the radiation emanating from them. He pioneered quantum theory by showing that the nuclei of atoms are surrounded by shells of electrons, each assigned particular sets of quantum numbers according to their orbits. He explained the

Danish physicists Niels Bohr and his son Aage Bohr. Both were awarded the Nobel Prize for Physics, with Niels Bohr being awarded the 1922 prize and Aage Bohr awarded the 1975 prize. AIP Emilio Segrè Visual Archives, Margrethe Bohr Collection

structure and behaviour of the nucleus, as well as the process of nuclear fission. Bohr also proposed the doctrine of **complementarity**, the theory that a fundamental particle is neither a wave nor a particle, because these are complementary modes of description.

Bohr's first model of the atom was developed while working in Manchester, UK, with Ernest Rutherford, who had proposed a nuclear theory of atomic structure from his work on the scattering of alpha rays in 1911. It was not, however, understood how electrons could continually orbit the nucleus without radiating energy, as classical physics demanded. In 1913, Bohr developed his theory of atomic structure by applying quantum theory to the observations of radiation emitted by atoms. Ten years earlier, Max Planck had proposed that radiation is emitted or absorbed by atoms in discrete units, or **quanta**, of energy. Bohr postulated that an atom may exist in only a certain number of stable states, each with a certain amount of energy, in which electrons orbit the nucleus without emitting or absorbing energy. He proposed that emission or absorption of energy occurs only with a transition from one stable state to another. When a transition occurs, an electron moving to a higher orbit absorbs energy and an electron moving to a lower orbit emits energy. In so doing, a set number of quanta of energy are emitted or absorbed at a particular frequency.

In 1939, Bohr proposed his liquid-droplet model for the nucleus, in which nuclear particles are pulled together by short-range forces, similar to the way in which molecules in a drop of liquid are attracted to one another. In the case of uranium, the extra energy produced by the absorption of a neutron causes the nuclear particles to separate into two groups of approximately the same size, thus breaking the nucleus into two smaller nuclei – a process called nuclear fission. The model was vindicated when Bohr correctly predicted the differing behaviour of nuclei of uranium-235 and uranium-238 from the fact that the numbers of neutrons in the two nuclei is odd and even respectively.

Bohr was born and educated in Copenhagen. In 1911, he went to the UK to study at the Cambridge atomic research laboratory under J J Thomson, but moved in 1912 to Manchester to work with Rutherford. He returned to Denmark as a lecturer at the University of Copenhagen in 1913, where he developed his theory of atomic structure. He moved back to Manchester to take up a lectureship offered by Rutherford, but the authorities in Denmark enticed him back, making him a professor in 1916 and then building the Institute of Theoretical Physics, where he was director from 1920. Leading physicists from all over the world developed Bohr's work at the institute, resulting in the more sophisticated quantum mechanics that more fully explained the behaviour of electrons and other elementary particles.

During World War II, he took part in work on the atomic bomb in the USA, after which he became a passionate advocate for the control of nuclear weapons. In 1952, Bohr was instrumental in creating the European Centre for Nuclear Research (CERN), now at Geneva, Switzerland. In addition to his scientific papers, Bohr published three volumes of essays: *Atomic Theory and the Description of Nature* (1934), *Atomic Physics and Human Knowledge* (1958), and *Essays 1958–1962 on Atomic Physics and Human Knowledge* (1963).

Born, Max (1882–1970)

German-born British physicist. He was awarded the Nobel Prize for Physics in 1954 for fundamental work on the quantum theory, especially his 1926 discovery that the wave function of an electron is linked to the probability that the electron is to be found at any point.

In 1924 Born coined the term 'quantum mechanics'. He made Göttingen a leading centre for theoretical physics and together with his students and collaborators – notably Werner Heisenberg – he devised in 1925 a system called matrix mechanics that accounted mathematically for the position and momentum of the electron in the atom. He also devised a technique, called the Born approximation method, for computing the behaviour of subatomic particles, which is of great use in high-energy physics.

Born was born in Breslau (now Wrocław, Poland) and studied at Breslau and Göttingen. He was professor of physics at Frankfurt-am-Main 1919–21 and at Göttingen 1921–33. With the rise to power of the Nazis, he left Germany for the UK, and in 1936 he became professor of natural philosophy at Edinburgh. In 1953 he retired to Germany.

Encouraged by German chemist Fritz Haber to study the lattice energies of crystals, in Frankfurt Born was able to determine the energies involved in lattice formation, from which the properties of crystals may be derived, and thus laid one of the foundations of solid-state physics.

He was inspired by Danish physicist Niels Bohr to seek a mathematical explanation for Bohr's discovery that the quantum theory applies to the behaviour of electrons in atoms. This led to matrix mechanics. But in 1926, Erwin Schrödinger expressed the same theory in terms of wave mechanics. Born used statistical probability to reconcile the two systems.

Bragg, (William) Lawrence (1890–1971)

Australian-born British physicist. He shared with his father William Bragg the Nobel Prize for Physics in 1915 for their research work on X-rays and crystals.

Bragg was born in Adelaide and studied mathematics there and at Cambridge, UK, then switched to physics. He was professor of physics at the University of Manchester 1919–38 and at Cambridge 1938–54.

He became interested in the work of Max von Laue, who claimed to have observed X-ray diffraction in crystals. Bragg was able to determine an equation now known as *Bragg's law* that enabled both him and his father to deduce the structure of crystals such as diamond, using the X-ray spectrometer built by his father. For this, he was the youngest person (at age 25) ever to receive the Nobel Prize. He was knighted in 1941.

Lawrence Bragg then went on to determine the structures of such inorganic substances as silicates.

Bragg, William Henry (1862–1942)

English physicist. He shared with his son Lawrence Bragg the Nobel Prize for Physics in 1915 for their research work on X-rays and crystals.

Crystallography had not previously been concerned with the internal arrangement of atoms but only with the shape and number of crystal surfaces. The Braggs' work gave a method of determining the positions of atoms in the lattices making up the crystals, and for accurate determination of X-ray wavelengths. This led to an understanding of the ways in which atoms combine with each other and revolutionized mineralogy and later molecular biology, in which X-ray diffraction was crucial to the elucidation of the structure of DNA.

Bragg was born in Westward, Cumberland. He obtained a first-class degree in mathematics from Cambridge in 1885 and was immediately appointed professor of mathematics and physics at the University of Adelaide, South Australia. In 1909 he returned to the UK as professor at Leeds; from 1915 he was professor at University College, London.

Bragg became convinced that X-rays behave as an electromagnetic wave motion. He constructed the first X-ray spectrometer in 1913. He and his son used it to determine the structures of various crystals on the basis that X-rays passing through the crystals are diffracted by the regular array of atoms within the crystal. He was knighted in 1920.

Bridgman, Percy Williams (1882–1961)

US physicist. His research into machinery producing high pressure led in 1955 to the creation of synthetic diamonds by General Electric. He was awarded the Nobel Prize for Physics in 1946 for his development of high-pressure physics.

Born in Cambridge, Massachusetts, he was educated at Harvard, where he spent his entire academic career.

Bridgman's experimental work on static high pressure began in 1908, and because this field of research had not been explored before, he had to invent much of his own equipment; for example, a seal in which the pressure in the gasket always exceeds that in the pressurized fluid. The result is that the closure is self-sealing. His discoveries included new, high-pressure forms of ice.

His technique for synthesizing diamonds was used to synthesize many more minerals and a new school of geology developed, based on experimental work at high pressure and temperature. Because the pressures and temperatures that Bridgman achieved simulated those deep below the ground, his discoveries gave an insight into the geophysical processes that take place within the Earth. His book *Physics of High Pressure* (1931) still remains a basic work.

Broglie, Louis Victor Pierre Raymond de (1892–1987)

7th duc de Broglie, French theoretical physicist. He established that all subatomic particles can be described either by particle equations or by wave equations, thus laying the foundations of wave mechanics. He was awarded the Nobel Prize for Physics in 1929 for his discovery of the wave-like nature of electrons. Succeeded as Duke in 1960.

De Broglie's discovery of wave–particle duality enabled physicists to view Einstein's conviction that matter and energy are interconvertible as being fundamental to the structure of matter. The study of matter waves led not only to a much deeper understanding of the nature of the atom but also to explanations of chemical bonds and the practical application of electron waves in electron microscopes.

De Broglie was born in Dieppe and educated at the Sorbonne, where he stayed on until 1928. He was professor at the Henri Poincaré Institute 1932–62. From 1946, he was a senior adviser on the development of atomic energy in France.

If particles could be described as waves, then they must satisfy a partial differential equation known as a wave equation. De Broglie developed such an equation in 1926, but found it in a form that did not offer useful information when it was solved.

Throughout his life, de Broglie was concerned with the philosophical issues of physics and he was the author of a number of books on this subject.

Chadwick, James (1891–1974)

English physicist. He was awarded the Nobel Prize for Physics in 1935 for his discovery in 1932 of the particle in the nucleus of an atom that became known as the *neutron* because it has no electric charge. He was knighted in 1945.

Chadwick established the equivalence of atomic number and atomic charge. During World War II, he was closely involved with the atomic bomb, and from 1943 he led the British team working on the Manhattan Project in the USA.

Chadwick was born in Cheshire and studied at Manchester under Ernest Rutherford, investigating the emission of gamma rays from radioactive materials. In 1913 he went to Berlin to work with German physicist Hans Geiger, inventor of the Geiger counter. There he was interned as an enemy alien during World War I. After the war, he joined Rutherford at Cambridge. Chadwick was professor of physics at Liverpool 1935–48 and master of Gonville and Caius College, Cambridge, 1948–58. Chadwick investigated beta particles emitted during radioactive decay. With Rutherford, he produced artificial disintegration of some of the lighter elements by alpha-particle bombardment. He returned to this at Liverpool, where he ordered the building of a cyclotron and developed a research school in nuclear physics.

Cockcroft, John Douglas (1897–1967)

British physicist. He was awarded the Nobel Prize for Physics in 1951 for his work in 1932 when, working with the Irish physicist Ernest Walton, with whom he shared the award, they succeeded in splitting the nucleus of an atom for the first time. He was knighted in 1948, and awarded the Order of Merit in 1957.

The voltage multiplier built by Cockcroft and Walton to accelerate protons was the first particle accelerator. They used it to bombard lithium, artificially transforming it into helium. The production of the helium nuclei was confirmed by observing their tracks in a cloud chamber. They then worked on the artificial disintegration of other elements, such as boron.

Cockcroft was born in Todmorden, West Yorkshire, and studied at Manchester and Cambridge, where he took up research work under Ernest Rutherford at the Cavendish Laboratory. Having been in charge of the construction of the first nuclear-power station in Canada during World War II, he returned to the UK to be director of Harwell Atomic Energy Research Establishment 1946–58, and in 1959 became first master of Churchill College, Cambridge.

Compton, Arthur H(olly) (1892–1962)

US physicist who in 1923 found that X-rays scattered by such light elements as carbon increased in wavelength. He concluded from this unexpected result that the X-rays were displaying both wavelike and particlelike properties, a phenomenon later named the *Compton Effect*. He was awarded the Nobel Prize for Physics in 1927 for his study of the transfer of energy

English nuclear physicist John Cockcroft. In the 1920s he played a leading role in the development of the particle accelerator, and in 1951 shared a Nobel prize with Ernest Walton for splitting the nucleus of an atom for the first time. AEA Technology

from electromagnetic radiation to a particle. He shared the award with Scottish physicist Charles Wilson. Compton was also a principal contributor to the development of the atomic bomb.

The behaviour of the X-ray, previously considered only as a wave, is explained best by considering that it acts as a corpuscle or particle of electromagnetic radiation – as a photon (Compton's term). Quantum mechanics benefited greatly from this interpretation. Further confirmation came from experiments using a cloud chamber in which collisions between X-rays and electrons were photographed and analysed.

Compton was born in Wooster, Ohio, studied at Princeton, and worked 1919–20 in the UK with nuclear physicist Ernest Rutherford at Cambridge. His academic career in the USA was spent at Washington University, St Louis, 1920–23 and 1945–61, and at the University of Chicago 1923–45. Compton became chancellor of Washington University, St Louis, Missouri (1945–54), and remained there until 1961. To determine whether cosmic rays consist of particles or electromagnetic radiation, Compton made measurements at various latitudes of comparative cosmic-ray intensities, using ionization chambers. By 1938 he had collated the results and demonstrated that the rays are deflected into curved paths by the Earth's magnetic field, proving that at least some components of cosmic rays consist of charged particles. During World War II, Chicago University was the prime location of the Manhattan Project, the effort to produce the first atomic bomb, and in 1942 Compton became one of its leaders. He organized research into methods of isolating fissionable plutonium and worked with Italian physicist Enrico Fermi on producing a self-sustaining nuclear chain reaction.

Cooper, Leon Niels (1930–　)

US physicist who in 1955 began work on the phenomenon of superconductivity. He proposed that at low temperatures electrons would be bound in pairs (since known as *Cooper pairs*) and in this state electrical resistance to their flow through solids would disappear. He was awarded the Nobel Prize for Physics in 1972 for his work on the theory of superconductivity. He shared the award with John Bardeen and J Robert Schrieffer.

Cooper was born in New York, where he attended Columbia University, specializing in quantum field theory – the interaction of particles and fields in subatomic systems. His work with John Bardeen was carried out at the University of Illinois. In 1958 Cooper moved to Brown University, Rhode Island, and in 1978 became director of the Centre for Neural Science at Brown.

Whereas the decrease in resistance is gradual in most metals, the resistance of superconductors suddenly disappears below a certain temperature. Experiments had shown that this temperature was inversely related to the mass of the nuclei. Cooper showed that an electron moving through the lattice attracts positive ions, slightly deforming the lattice. This leads to a momentary concentration of positive charge that attracts a second electron. This is the Cooper pair. Although the electrons in the pair are only weakly bound to each other, Bardeen, Cooper, and Schrieffer were able to show that they all form a single quantum state with a single momentum. The scattering of individual electrons does not affect this momentum, and this leads to zero resistance. Cooper pairs cannot be formed above the critical temperature, and superconductivity breaks down.

Developing a theory of the central nervous system, Cooper has also worked on distributed memory and character recognition.

He published *The Meaning and Structure of Physics* (1968).

Curie, Marie (1867–1934)

Polish scientist, born *Maria Sklodowska,* who, with husband Pierre Curie, discovered in 1898 two new radioactive elements in pitchblende ores: polonium and radium. They isolated the pure elements in 1902. Both scientists refused to take out a patent on their discovery and were jointly awarded the Nobel Prize for Physics in 1903, with Henri Becquerel, for their research on radiation phenomena. Marie Curie was also awarded the Nobel Prize for Chemistry in 1911 for the discovery of radium and polonium, and the isolation and study of radium.

From 1896 the Curies worked together on radioactivity, building on the results of Wilhelm Röntgen (who had discovered X-rays) and Becquerel (who had discovered that similar rays are emitted by uranium salts). Marie Curie discovered that thorium emits radiation and found that the mineral pitchblende was even more radioactive than could be accounted for by any uranium and thorium content. In July 1898, the Curies announced the discovery of polonium, followed by the discovery of radium five months later. They eventually prepared 1 g/0.04 oz of pure radium chloride – from 8 tonnes of waste pitchblende from Austria.

They also established that beta rays (now known to consist of electrons) are negatively charged particles. In 1910 with André Debierne (1874–1949), who had discovered actinium in pitchblende in 1899, Marie Curie isolated pure radium metal in 1911.

Maria Sklodowska was born in Warsaw, then under Russian domination. She studied in Paris from 1891 and married in 1895. In 1906, after her husband's death, she succeeded him as professor of physics at the Sorbonne; she was the first woman to teach there. She wrote a *Treatise on Radioactivity* (1910).

At the outbreak of World War I in 1914, Curie helped to install X-ray equipment in ambulances, which she drove to the front lines. The International Red Cross made her head of its Radiological Service. Assisted by daughter Irene and Martha Klein at the Radium Institute, she held courses for medical orderlies and doctors, teaching them how to use the new technique.

By the late 1920s her health began to deteriorate: continued exposure to high-energy radiation had given her leukaemia. She and her husband had taken no precautions against radioactivity. Her notebooks, even today, are too contaminated to handle. She entered a sanatorium at Haute Savoie and died there in July 1934, a few months after her daughter and son-in-law, the Joliot-Curies, had announced the discovery of artificial radioactivity.

Davisson, Clinton Joseph (1881–1958)

US physicist who in 1927 made the first experimental observation of the wave nature of electrons, for which he was awarded the Nobel Prize for Physics in 1937. He shared the award with George Thomson who carried through the same research independently.

Davisson was born in Bloomington, Illinois, and studied at the University of Chicago. He worked for the Western Electric Company (later Bell Telephone) in New York 1917–46. With Lester Germer (1896–1971), Davisson discovered that electrons can undergo diffraction, in accordance with French physicist Louis de Broglie's theory that electrons and all other elementary particles can show wavelike behaviour.

Davisson and Germer used a single nickel crystal in their experiments. The atoms were in a cubic lattice with atoms at the apex of cubes, and the electrons were directed at the plane of atoms at 45° to the regular and plane. Electrons of a known velocity were directed at this plane and those emitted were detected by a Faraday chamber. In January 1927, results showed that at a certain velocity of incident electrons, diffraction occurred, producing outgoing beams that could be related to the interplanar distance. The wavelength of the beams was determined, and this was then used with the known velocity of the elctrons to verify de Broglie's hypothesis. The first work gave results with an error of 1–2% but later systematic work produced results in complete agreement. (Similar experiments at higher voltages using metal foil were carried out later in the same year at Aberdeen University, UK, by George Thomson, the son of J J Thomson). The particle–wave duality of subatomic particles was established beyond doubt, and both men were awarded the 1937 Nobel Prize for Physics.

De Forest, Lee (1873–1961)

US physicist and inventor who in 1906 invented the triode valve, which contributed to the development of radio, radar, and television.

In 1904 Ambrose Fleming invented the diode valve. De Forest saw that if a third electrode were added, the triode valve would serve as an amplifier as well as a rectifier, and radio communications would become a practical possibility.

De Forest was born in Council Bluffs, Iowa, and studied at Yale. Working for the Western Electric Company in Chicago, he devised ways of rapidly transmitting wireless signals, his system being used in 1904 in the first wireless news report (of the Russo-Japanese War).

De Forest set up his own wireless telegraph company, but nearly went bankrupt twice. He was prosecuted for attempting to use the US mail to defraud, by seeking to promote the 'worthless' audion tube (as he called the triode valve).

In 1912, De Forest arranged triode valves to transmit both speech and music by radio, and in 1916 he set up a radio station and began broadcasting news.

In 1923, De Forest demonstrated an early system of motion pictures carrying a soundtrack, called phonofilm. Its poor quality, and lack of interest from film-makers, led to its demise, though the principle was later adopted.

Dirac, Paul Adrien Maurice (1902–1984)

English physicist who worked out a version of quantum mechanics consistent with special relativity. The existence of antiparticles, such as the positron (positive electron), was one of its predictions. He shared the Nobel Prize for Physics in 1933 (with Austrian physicist Erwin Schrödinger) for his work on the development of quantum mechanics.

Dirac was born and educated in Bristol and from 1923 at Cambridge, where he was professor of mathematics 1932–69. From 1971 he was professor of physics at Florida State University. In 1928 Dirac formulated the relativistic theory of the electron. The model was able to describe many quantitative aspects of the electron, including such properties as quantum spin. Dirac noticed that those particles with half-integral spins obeyed statistical rules different from the other particles. For these particles, Dirac worked out the statistics, now called *Fermi–Dirac statistics* because Italian physicist Enrico Fermi had done very similar work. These are used, for example, to determine the distribution of electrons at different energy levels.

Einstein, Albert (1879–1955)

German-born US physicist whose theories of relativity revolutionized our understanding of matter, space, and time. Einstein established that light may have a particle nature. He was awarded the Nobel Prize for Physics in 1921 for his work on theoretical physics, especially the *photoelectric law*. He also investigated Brownian motion, confirming the existence of atoms. His last conception of the basic laws governing the universe was outlined in his unified field theory, made public in 1953.

Einstein's first major achievement concerned Brownian movement, the random movement of fine particles that can be seen through a microscope, which was first observed in 1827 by Robert Brown when studying a suspension of pollen grains in water. The motion of the pollen grains increased when the temperature increased but decreased if larger particles were used. Einstein explained this phenomenon as being the effect of large numbers of molecules (in this case, water molecules) bombarding the particles. He was able to make predictions of the movement and sizes of the particles, which were later verified experimentally by the French physicist Jean Perrin. Einstein's explanation of Brownian motion and its subsequent experimental confirmation was one of the most important pieces of evidence for the hypothesis that matter is composed of atoms. Experiments based on this work were used to obtain an accurate value of Avogadro's number (the number of atoms in one mole of a substance) and the first accurate values of atomic size.

Einstein's work on photoelectricity began with an explanation of the radiation law proposed in 1901 by Max Planck: $E = h\nu$, where E is the energy of radiation, h is Planck's constant, and ν is the frequency of radiation. Einstein suggested that packets of light energy are capable of behaving as particles called 'light quanta' (later called photons). Einstein used this hypothesis to explain the photoelectric effect, proposing that light particles striking the surface of certain metals cause electrons to be emitted. It had been found experimentally that electrons are not emitted by light of less than a certain frequency ν^0; that when electrons are emitted, their energy increases with an increase in the frequency of the light; and that an increase in light intensity produces more electrons but does not increase their energy. Einstein suggested that the kinetic energy of each electron, $\frac{1}{2} m\nu^2$, is equal to the difference in the incident light energy, $h\nu$, and the light energy needed to overcome the threshold of emission, $h\nu^0$. This can be written mathematically as: $\frac{1}{2} m\nu^2 = h\nu - h\nu^0$

The *special theory of relativity* started with the premises that (1) the laws of nature are the same for all observers in unaccelerated motion, and (2) the speed of light is independent of the motion of its source. Until then, there had been a steady accumulation of knowledge that suggested that light and other electromagnetic radiation does not behave as predicted by classical physics. For example, various experiments, including the *Michelson–Morley experiment,* failed to measure the expected changes in the speed of light relative to the motion of the Earth. Such experiments are now interpreted as showing that no 'ether' exists in the universe as a medium to carry light waves, as was required by classical physics. Einstein recognized that light has a measured speed that is independent of the speed of the observer. Thus, contrary to everyday experience with phenomena such as sound waves, the velocity of light is the same for an observer travelling at high speed *towards* a light source as it is for an observer travelling rapidly *away* from the light source. To Einstein it followed that, if the speed of light is the same for both these observers, the time and distance framework they use to measure the speed of light cannot be the same. Time and distance vary, depending on the velocity of each observer. From the notions of relative motion and the constant velocity of light, Einstein derived the result that, in a system in motion relative to an observer, length would be observed to decrease, time would slow down, and mass would increase. The magnitude of these effects is negligible at ordinary velocities and Newton's laws still hold good. But at velocities approaching that of light, they become substantial. As a system approaches the velocity of light, relative to an observer at rest, its length decreases towards zero, time slows almost to a stop, and its mass increases without limit. Einstein therefore concluded that no system can be accelerated to a velocity equal to or greater than the velocity of light. Einstein's conclusions regarding time dilation and mass increase were verified with observations of fast-moving atomic clocks and cosmic rays. Einstein showed in 1907 that mass is related to energy by the famous equation $E=mc^2$, which indicates the enormous amount of energy that is stored as mass, some of which is released in radioactivity and nuclear reactions, for example in the Sun.

In the ***general theory of relativity*** (1916), the properties of space–time were to be conceived as modified locally by the presence of a body with mass; and light rays should bend when they pass by a massive object. A planet's orbit around the Sun arises from its natural trajectory in modified space–time. General relativity theory was inspired by the simple idea that it is impossible in a small region to distinguish between acceleration and gravitation effects (as in a lift one feels heavier when it accelerates upwards). Einstein used the general theory to account for an anomaly in the orbit of the planet Mercury that could not be explained by Newtonian mechanics. Furthermore, the general theory made two predictions concerning light and gravitation. The first was that a red shift is produced if light passes through an intense gravitational field, and this was subsequently detected in astronomical observations in 1925. The second was a prediction that the apparent positions of stars would shift when they are seen near the Sun because the Sun's intense gravity would bend the light rays from the stars as they pass the Sun. Einstein was triumphantly vindicated when observations of a solar eclipse in 1919 showed apparent shifts of exactly the amount he had predicted.

Einstein was born in Ulm, Württemberg, and lived in Munich and Italy before settling in Switzerland. Disapproving of German militarism, he became a Swiss citizen in 1901 and was appointed an inspector of patents in Berne in 1902. In his spare time, he took his PhD at Zürich. In 1909, he became a lecturer in theoretical physics at the university. After holding a similar post at Prague in 1911, he returned to teach at Zürich in 1912. In 1913, he took up a specially created post as director of the Kaiser Wilhelm Institute for Physics, Berlin. Confirmation of the general theory of relativity by the solar eclipse of 1919 made Einstein world famous. After the Nazis came to power, he emigrated to the USA in 1933 and became professor of mathematics and a permanent member of the Institute for Advanced Study at Princeton, New Jersey. In 1939, Einstein drew the attention of the president of the USA to the possibility that Germany might be developing the atomic bomb. This prompted US efforts to produce the bomb, although Einstein did not take part in them. When the US bomb was used, he was reported as saying that, if he had known his theories would lead to such destruction, he would rather have been a watchmaker. After World War II, Einstein was actively involved in the movement to abolish nuclear weapons. In 1952, the state of Israel paid him the highest honour it could by offering him the presidency, which he declined.

Esaki, Leo (1925–   )

Japanese physicist who shared the 1973 Nobel Prize for his discovery of tunnelling in semiconductor diodes. These devices are now called Esaki diodes. Esaki spent most of his working life in the USA but returned to Japan in the 1990s as president of the University of Tsukuba.

Esaki was born on 12 March 1925 in Osaka. When he graduated from the University of Tokyo in 1947, he wanted to become a nuclear physicist but Japan lacked the particle accelerators needed to compete in this field and he moved to solid-state physics. Working for Sony in 1957, Esaki noticed while studying very small heavily-doped germanium diodes that sometimes, and unexpectedly, the resistance decreased as current increased. This was caused by 'tunnelling' – a quantum mechanical effect whereby electrons can travel (tunnel) through electrostatic potentials that they would be unable to overcome classically. These barriers have to be very thin for tunnelling to occur and Esaki was able to use this effect for switching and to build ultra-small and ultra-fast tunnel diodes. In 1959 Esaki received a PhD from Tokyo for this discovery. It also earned Esaki the 1973 Nobel Prize, which he shared with Brian Josephson, who predicted supercurrents in superconducting diodes, and Ivar Giaever who discovered electron tunnelling in superconductors. In 1960 Esaki joined IBM's Thomas J Watson Research Center in Yorktown Heights, New York, where he became an IBM fellow, the company's highest research honour, in 1967. He continued to research the nonlinear transport and optical properties of semiconductors, in particular multilayer superlattice structures grown by molecular beam epitaxy techniques. In 1992 Esaki returned to Japan to become president of Tsukuba University.

The return of the nation's only living physics Nobel laureate was front-page news in Japan.

Fermi, Enrico (1901–1954)

Italian-born US physicist who was awarded the Nobel Prize for Physics in 1938 for his proof of the existence of new radioactive elements produced by bombardment with neutrons, and his discovery of nuclear reactions produced by low-energy neutrons. This research was the basis for studies leading to the atomic bomb and nuclear energy. Fermi built the first nuclear reactor in 1942 at Chicago University and later took part in the Manhattan Project to construct an atom bomb. His theoretical work included the study of the weak nuclear force, one of the fundamental forces of nature, and beta decay.

Following the work of the Joliot-Curies, who discovered artificial radioactivity in 1934 using alpha particle bombardment, Fermi began producing new radioactive isotopes by neutron bombardment. Unlike the alpha particle, which is positively charged, the neutron is uncharged. Fermi realized that less energy would be wasted when a bombarding neutron encounters a positively charged target nucleus. He also found that a block of paraffin wax or a jacket of water around the neutron source produced slow, or 'thermal', neutrons. Slow neutrons are more effective at producing artificial radioactive elements because they remain longer near the target nucleus and have a greater chance of being absorbed. He did, however, misinterpret the results of experiments involving neutron bombardment of uranium, failing to recognize that nuclear fission had occurred. Instead, he maintained that the bombardment produced two new transuranic elements. It was left to Lise Meitner and Otto Frisch to explain nuclear fission in 1938.

In the USA, Fermi continued the work on the fission of uranium (initiated by neutrons) by building the first nuclear reactor, then called an *atomic pile*, because it had a moderator consisting of a pile of purified graphite blocks (to slow the neutrons) with holes drilled in them to take rods of enriched uranium. Other neutron-absorbing rods of cadmium, called control rods, could be lowered into or withdrawn from the pile to limit the number of slow neutrons available to initiate the fission of uranium. The reactor was built on the squash court of Chicago University. On the afternoon of 2 December 1942, the control rods were withdrawn for the first time and a self-sustaining nuclear chain reaction began. Two years later, the USA, through a team led by Arthur Compton and Fermi, had constructed an atomic bomb, in which the same reaction occurred but was uncontrolled, resulting in a nuclear explosion.

Fermi's experimental work on beta decay in radioactive materials provided further evidence for the existence of the neutrino, predicted by Austrian physicist Wolfgang Pauli.

Fermi was born in Rome and studied at Pisa; Göttingen, Germany; and Leiden, the Netherlands. He was professor of theoretical physics at Rome 1926–38, where he wrote *Introduzione alla Fisica Atomica* (1928), the first textbook on modern physics to be published in Italy. The rise of fascism in Italy caused him to emigrate to the USA. He was professor at Columbia University, New York, 1939–42. At the end of World War II, Fermi became a US citizen in 1945 and returned to Chicago to continue his researches as professor of physics. With British physicist Paul Dirac, Fermi studied the quantum statistics of particles with half-integer spin, which are named fermions after him.

Feynman, Richard P(hillips) (1918–1988)

US physicist whose work laid the foundations of quantum electrodynamics. He was awarded the Nobel Prize for Physics in 1965 for his work on the theory of radiation. He shared the award with Julian Schwinger and Sin-Itiro Tomonaga. He also contributed to many aspects of particle physics, including quark theory and the nature of the weak nuclear force.

For his work on quantum electrodynamics, he developed a simple and elegant system of *Feynman diagrams* to represent interactions between particles and how they moved from one space-time point to another. He derived rules for calculating the probability of the interaction represented by each diagram. His other major discoveries are the theory of superfluidity (frictionless flow) in liquid helium, developed in the early 1950s; his work on the weak interaction (with US physicist Murray Gell-Mann) and the strong force; and his prediction that the proton and neutron are not elementary particles. Both particles are now known to be composed of quarks.

Feynman was born in New York and studied at the Massachusetts Institute of Technology and at Princeton. During World War II, he worked at Los Alamos, New Mexico, on the behaviour of neutrons in atomic explosions. Feynman was professor of theoretical physics at Caltech (California Institute of Technology) from 1950 until his death. As a member of the committee investigating the *Challenger* space-shuttle disaster 1986, he demonstrated the faults in rubber seals on the shuttle's booster rocket. *The Feynman Lectures on Physics* (1963) became a standard work. He also published two volumes of autobiography: *Surely You're Joking, Mr Feynman!* (1985), and *What Do You Care What Other People Think?* (1988).

Franck, James (1882–1964)

German-born US physicist. He shared the Nobel Prize for Physics in 1925 with his co-worker Gustav Hertz for their experiments of 1914 on the energy transferred by colliding electrons to mercury atoms, showing that the transfer was governed by the rules of quantum theory.

Franck was born in Hamburg and educated at Heidelberg and Berlin. In 1920 he became professor of experimental physics at Göttingen, but emigrated to the USA in 1933 after publicly protesting against the Nazis' racial policies. He was a professor at the University of Chicago 1938–49. He participated in the wartime atomic-bomb project at Los Alamos but organized the 'Franck petition' in 1945, which argued that the bomb should not be used against Japanese cities. After World War II he turned his research to photosynthesis.

Investigating the collisions of electrons with rare-gas atoms, Franck found that they are almost completely elastic and that no kinetic energy is lost. With Hertz, he extended this work to other atoms. This led to the discovery that there are inelastic collisions in which energy is transferred in definite amounts.

Franck also studied the formation, dissociation, vibration, and rotation of molecules, and was able to calculate the dissociation energies of molecules. Edward Condon interpreted this method in terms of wave mechanics, and it has become known as the *Franck–Condon principle.*

Frisch, Otto Robert (1904–1979)

Austrian-born British physicist who first described the fission of uranium nuclei under neutron bombardment, coining the term 'fission' to describe the splitting of a nucleus.

Frisch was born and educated in Vienna. Doing research at Hamburg, he fled from Nazi Germany in 1933, initially to the UK and in 1934 to the Institute of Theoretical Physics in Copenhagen. The German occupation of Denmark at the beginning of World War II forced Frisch to return to Britain. He then worked 1943–45 on the atom bomb at Los Alamos, New Mexico, USA. He was professor of natural philosophy at Cambridge University 1947–71. Frisch worked on methods of separating the rare uranium-235 isotope that would undergo fission. He also calculated details such as the critical mass needed to produce a chain reaction and make an atomic bomb, and urged the British government to undertake nuclear research. At the first test explosion of the atomic bomb, Frisch conducted experiments from a distance of 40 km/25 mi. He was the nephew of physicist Lise Meitner.

Gabor, Dennis (1900–1979)

Hungarian-born British physicist. He was awarded the Nobel Prize for Physics in 1971 for his invention in 1947 of the holographic method of three-dimensional photography.

Born in Budapest, Gabor studied at the Budapest Technical University and then at the Technishe Hochschule in Berlin. He worked in Germany until he fled to Britain in 1933 to escape the Nazis. From 1958 to 1967 he was professor of applied electron physics at the Imperial College of Science and Technology, London.

When Gabor began work on holography, he considered the possibility of improving the resolving power of the electron microscope, first by using the electron beam to make a hologram of the object and then by examining this hologram with a beam of coherent light. But coherent light of sufficient intensity was not achievable until the laser was demonstrated in 1960.

Other work included research on high-speed oscilloscopes, communication theory, and physical optics. In 1958 he invented a type of colour television tube of greatly reduced depth. He took out more than 100 patents for his inventions.

Gamow, George (1904–1968)

Russian-born US cosmologist, nuclear physicist, and popularizer of science, born *Georgi Antonovich Gamow*. His work in astrophysics included a study of the structure and evolution of stars and the creation of the elements. He explained how the collision of nuclei in the solar interior could produce the nuclear reactions that power the Sun. With the 'hot Big Bang' theory, which he co-proposed in 1948, he indicated the origin of the universe.

Gamow predicted that the electromagnetic radiation left over from the universe's formation should, after having cooled down during the subsequent expansion of the universe, manifest itself as a microwave cosmic background radiation. He also made an important contribution to the understanding of protein synthesis.

Gamow was born in Odessa (now in Ukraine), and studied at the universities of Leningrad (St Petersburg) and Göttingen, Germany. He then worked at the Institute of Theoretical Physics in Copenhagen, Denmark, and at the Cavendish Laboratory, Cambridge, UK. From 1931 to 1933 he was master of research at the Academy of Science in Leningrad, and then defected to the USA, becoming professor of physics at George Washington University in Washington, DC, (1934–56) and then at the University of Colorado (1956–68). In the late 1940s, he worked on the hydrogen bomb at Los Alamos, New Mexico.

Gamow's model of alpha decay (1928) represented the first application of quantum mechanics to the study of nuclear structure. Later he described beta decay (see radioactive decay).

With US physicist Ralph Alpher and German-born US physicist Hans Bethe, he investigated the possibility that heavy elements could have been produced by a sequence of neutron-capture thermonuclear reactions. They published a paper in 1948, which became known as the Alpher–Bethe–Gamow (or alpha–beta–gamma) hypothesis, describing the 'hot Big Bang'.

Gamow also contributed to the solution of the genetic code. The double-helix model for the structure of DNA involves four types of nucleotides (complex organic compounds). Gamow realized that if three nucleotides were used at a time, the possible combinations could easily code for the different amino acids of which all proteins are constructed. Gamow's theory was found to be correct in 1961.

Geiger, Hans (Wilhelm) (1882–1945)

German physicist who produced the Geiger counter. He spent the period 1906–12 in Manchester, UK, working with Ernest Rutherford on radioactivity. In 1908 they designed an instrument to detect and count alpha particles, positively charged ionizing particles produced by radioactive decay.

In 1928 Geiger and Walther Müller produced a more sensitive version of the counter, which could detect all kinds of ionizing radiation.

Geiger was born in Neustadt, Rheinland-Pfalz, and studied at Munich and Erlangen. On his return from Manchester in 1912, Geiger became head of the Radioactivity Laboratories at the Physikalische Technische Reichsanstalt in Berlin, where he established a successful research group. He subsequently held other academic posts in Germany; from 1936 he was at the Technical University, Berlin. Subjects that Geiger studied under Rutherford include the mathematical relationship between the amount of alpha scattering and atomic weight, the relationship between the range of an alpha particle and its velocity, the various disintegration products of uranium, and the relationship between the range of an alpha particle and the decay constant, which is a measure of the rate at which alpha particles are emitted. From 1931 onwards Geiger mainly studied cosmic radiation.

Gell-Mann, Murray (1929–)

US theoretical physicist. In 1964 he formulated the theory of the quark as one of the fundamental constituents of matter. He was awarded the Nobel

Prize for Physics in 1969 for his work on elementary particles and their interaction.

In 1962 Gell-Mann proposed a classification system for elementary particles called the *eightfold way*. It postulated the existence of supermultiplets, or groups of eight particles which have the same spin value but different values for charge, isotopic spin, mass, and strangeness. The model also predicted the existence of supermultiplets of different sizes.

Gell-Mann was born in New York and studied at Yale and the Massachusetts Institute of Technology. He became professor at the California Institute of Technology in 1956.

Gell-Mann proposed in 1953 a new quantum number called the strangeness number, together with the law of conservation of strangeness, which states that the total strangeness must be conserved on both sides of an equation describing a strong or an electromagnetic interaction but *not* a weak interaction. This led to his theory of associated production in 1955 concerning the creation of strange particles. Gell-Mann used these rules to group mesons, nucleons (neutrons and protons), and hyperons, and was thereby able to form successful predictions.

Glashow, Sheldon Lee (1932–)

US particle physicist who proposed the existence of a fourth 'charmed' quark and later argued that quarks must be coloured. Insights gained from his theoretical studies enabled him to consider ways in which the weak nuclear force and the electromagnetic force could be unified as a single force (now called the electroweak force). He shared the 1979 Nobel Prize for Physics with Pakistani physicist Abdus Salam and US physicist Steven Weinberg for this work.

Sheldon Glashow was born in Manhattan on 5 December 1932, the son of immigrant Jewish parents from Tsarist Russia and the third of three brothers. After years of struggle his father became a plumber and was able to provide for the university education for his sons that he and his wife had missed. Sheldon's brothers became a dentist and a doctor but Sheldon wanted to be a scientist from an early age, and at the age of 15 he had a basement laboratory in which he carried out dangerous research on the synthesis of selenium halides. His parents always encouraged him in his scientific pursuits, only occasionally suggesting that he should take the safe route of becoming a doctor first.

Glashow was educated at the Bronx High School of Science where Weinberg was his classmate in a highly talented class. In 1950 he entered Cornell University (with Weinberg) obtaining a bachelor's degree in 1954 and moving to Harvard to gain an MA in 1955 and a PhD in 1958. His thesis at Harvard was called 'The vector meson in elementary particle decays', reflecting an early interest in electroweak synthesis.

Winning a postdoctoral fellowship, Glashow's intention was to work in Moscow but his visa never came, so he spent his fellowship 1958–60 at the Niels Bohr Institute in Copenhagen where he discovered the SU(2) × U(I) structure of the electroweak theory and also later worked briefly, in 1964, with US physicist James D Bjorken (1934–) on a property that he called 'charm'. His research took him to the European Organization for Nuclear Research (CERN) in Geneva, Switzerland, and then to the California Institute of Technology (Caltech) in Stanford where he spent several years working with Sidney Coleman on the eightfold way invented by Murray Gell-Mann. He was a member of the faculty at Berkeley 1962–66, returning to Harvard as professor of physics where he remained, except for periods spent at CERN (1968), Marseilles (1970), MIT (1974), and Texas (1982). In July 2000 he took up a teaching post at Boston University, Massachusetts.

John Iliopoulos and Luciano Miani arrived at Harvard in 1969 as research fellows and joined Glashow in predicting the existence of 'charmed hadrons'. He later collaborated with Alvaro de Rujula and Howard Georgi in correctly predicting the presence of what he called 'charm' in neutrino physics or in e + e = annihilation, thus realizing that many strands of research were converging on one theory of physics.

Glashow shared the Nobel Prize for Physics with Steven Weinberg and Abdus Salam in 1979 for what was cited as 'their contribution to the theory of the unified weak and electromagneticinteraction between elementary particles, including *inter alia* the prediction of the weak neutral current'. Working on Glashow's early model explaining the electromagnetic and weak nuclear forces, Weinberg and Salam developed a coherent theory that unified two of the four forces of physics: the electromagnetic interaction and the weak interaction, which they applied to leptons (electrons and neutrinos). Using Gell-Mann's theory that particles are made up of smaller particles called quarks, Glashow then postulated that a fourth quark was necessary and extended the Weinberg–Salam theory to other particles – baryons and mesons – by introducing the particle property 'charm'. Glashow predicted the existence of other new particles, which were found in the following years, and the quark theory was expanded to include a coloured quark. The theory of quantum chromodynamics is based on his approach.

Hahn, Otto (1879–1968)

German physical chemist who was awarded the Nobel Prize for Chemistry in 1944 for his discovery of nuclear fission (see nuclear energy). In 1938 with Fritz Strassmann (1902–1980), he discovered that uranium nuclei split when bombarded with neutrons. Hahn did not participate in the resultant development of the atom bomb.

In 1918, Hahn and Lise Meitner discovered the longest-lived isotope of a new element which they called protactinium, and in 1921 they discovered nuclear isomers – radioisotopes with nuclei containing the same subatomic particles but differing in energy content and half-life.

Hahn was born in Frankfurt-am-Main and studied at Marburg. From 1904 to 1906 he worked in London under William Ramsay, who introduced Hahn to radiochemistry, and at McGill University in Montréal, Canada, with Ernest Rutherford. Returning to Germany, Hahn was joined at Berlin in 1907 by Meitner, beginning a long collaboration. Hahn was director of the Kaiser Wilhelm Institute for Chemistry in Berlin 1928–44, and then president of the Max Planck Institute in Göttingen.

Hawking, Stephen (William) (1942–   )

English physicist whose work in general relativity – particularly gravitational field theory – led to a search for a quantum theory of gravity to explain black holes and the Big Bang, singularities that classical relativity theory does not adequately explain. His book *A Brief History of Time* (1988) gives a popular account of cosmology and became an international bestseller. He later co-wrote (with Roger Penrose) *The Nature of Space and Time* (1996).

Hawking's objective of producing an overall synthesis of quantum mechanics and relativity theory began around the time of the publication in 1973 of his seminal book *The Large Scale Structure of Space-Time*, written with G F R Ellis. His most remarkable result, published in 1974, was that black holes could in fact emit particles in the form of thermal radiation – the so-called *Hawking radiation*.

Hawking was born in Oxford, studied at Oxford and Cambridge, and became professor of mathematics at Cambridge in 1979. He developed motor neurone disease, a rare form of degenerative paralysis, while still a college student, and uses a wheelchair and communication equipment. Hawking's most fruitful work was with black holes, stars that have undergone total gravitational collapse and whose gravity is so great that nothing, not even light, can escape from them. Since 1974, he has studied the behaviour of matter in the immediate vicinity of a black hole. He has proposed a physical explanation for Hawking radiation which relies on the quantum-mechanical concept of 'virtual particles' – these exist as particle–antiparticle pairs and fill 'empty' space. Hawking suggested that, when such a pair is created near a black hole, one half of the pair might disappear into the black hole, leaving the other half, which could escape to infinity. This would be seen by a distant observer as thermal radiation.

English physicist Stephen Hawking, professor of mathematics at Cambridge University, UK. Hawking is a leading researcher into black holes and gravitational field theory.
AIP Emilio Segrè Visual Archives, Physics Today Collection

Heisenberg, Werner (Karl) (1901–1976)

German physicist who developed quantum theory and formulated the uncertainty principle, which places absolute limits on the achievable accuracy of measurement. He was awarded the Nobel Prize for Physics in 1932 for his creation of quantum mechanics, work he carried out when only 24.

Heisenberg was concerned not to try to picture what happens inside the atom but to find a mathematical system that explained it. His starting point was the spectral lines given by hydrogen, the simplest atom. Assisted by Max Born, Heisenberg presented his ideas in 1925 as a system called *matrix mechanics*. He obtained the frequencies of the lines in the hydrogen spectrum by mathematical treatment of values within matrices or arrays. His work was the first precise mathematical description of the workings of the atom and with it Heisenberg is regarded as founding quantum mechanics, which seeks to explain atomic structure in mathematical terms. Heisenberg also was able to predict from studies of the hydrogen spectrum that hydrogen exists in two allotropes – ortho-hydrogen and para-hydrogen – in which the two nuclei of the atoms in a hydrogen molecule spin in the same or opposite directions respectively. The allotropes were discovered in 1929. In 1927 Heisenberg made the discovery of the *uncertainty principle*, for which he is best known. The uncertainty principle states that there is a theoretical limit to the precision with which a particle's position and momentum can be measured. In other words, it is impossible to specify precisely both the position and the simultaneous momentum (mass multiplied by velocity) of a particle. There is always a degree of uncertainty in either, and as one is determined with greater precision, the other can only be found less exactly. Multiplying together the uncertainties of the position and momentum yields a value approximately equal to Planck's constant. The idea that the result of an action can be expressed only in terms of the probability of a certain effect was revolutionary, and it discomforted even Albert Einstein, but is generally accepted today. In 1927 Heisenberg used the Pauli exclusion principle, which states that no two electrons can have identical sets of quantum numbers the same, to show that ferromagnetism (the ability of some materials to acquire magnetism in the presence of an external magnetic field) is caused by electrostatic interaction between the electrons.

Heisenberg was born in Würzburg and studied at Munich. In 1923 he became assistant to Max Born at Göttingen, then worked with Danish physicist Niels Bohr in Copenhagen 1924–26. Heisenberg became professor at Leipzig in 1927 at the age of only 26, and stayed until 1941. He was director of the Max Planck Institute for Physics 1942–70. During World War II, Heisenberg worked for the Nazis on nuclear fission, but his team was many months behind the Allied atom-bomb project. After the war he worked on superconductivity.

Hertz, Gustav (1887–1975)

German physicist who, with US physicist James Franck, demonstrated that mercury atoms, when bombarded with electrons, absorb energy in discrete units (or quanta). Following the absorption of energy, the atoms return to their original state by emitting a photon of light. This was the first experimental proof that the quantum theory of atoms was correct and demonstrated the reality of atomic energy levels. He shared the Nobel Prize for Physics in 1925 with James Frank for the discovery of the laws governing the impact of an electron upon an atom.

Gustav Hertz was born in Hamburg, Germany, and obtained his doctorate in Berlin in 1911. He became an assistant in the Berlin physics institute and began his collaboration with Franck. He was professor of experimental physics at the University of Halle 1925–1927. From 1928 to 1935 he worked at the Berlin Techniche Hochschule. Because of his Jewish descent, Hertz was forced to resign in 1935. However, he remained in Germany during World War II, working as director of the Siemens Research Laboratory, Berlin. After the war, he was captured by the Russians and taken to the USSR to continue his work in atomic physics. In 1955 he re-emerged as director of the Physics Institute in Leipzig, East Germany. He is the nephew of Heinrich Hertz, the discoverer of radio waves.

Higgs, Peter Ware (1929–)

English theoretical physicist who is best known for his prediction of an as yet undiscovered fundamental particle known as the *Higgs boson*. The Higgs boson, if discovered, will lead to our understanding of the origin of mass.

Born in Bristol, Higgs attended school in Birmingham before studying for his BSc in physics at King's College, London. Higgs remained at King's College and gained his PhD in 1955. After a brief spell at the University of London, Higgs moved to the University of Edinburgh in 1960, becoming professor of theoretical physics in 1980. His many prizes include the 1997 Paul Dirac medal and the 1998 European Physical Society High Energy and Particle Physics Prize.

Josephson, Brian David (1940–)

Welsh physicist, a leading authority on superconductivity. He shared the Nobel Prize for Physics in 1973 for his theoretical predictions of the properties of a supercurrent through a tunnel barrier (the *Josephson effect*), which led to the development of the Josephson junction.

Josephson was born in Cardiff and studied at Cambridge University, where he has spent most of his career, becoming professor in 1974. In 1962, Josephson saw some novel connections between solid-state theory and his own experimental problems in superconductivity. He then calculated the current due to quantum mechanical tunnelling across a thin strip of insulator between two superconductors, and the current–voltage characteristics of such junctions are now known as the Josephson effect. The Josephson effect may be used as a generator of radiation, particularly in the microwave and far infrared region, and in detecting tiny anomalies in a magnetic field.

Kapitza, Peter Leonidovich (1894–1984)

Soviet physicist, also known as *Pyotr Kapitsa,* who shared the Nobel Prize for Physics in 1978 for his work on magnetism and low-temperature physics. He worked on the superfluidity of liquid helium and also achieved the first high-intensity magnetic fields.

Kapitza was born near St Petersburg and studied at Petrograd Polytechnical Institute, after which he went to the UK and worked at the Cavendish Laboratory, Cambridge, with nuclear physicist Ernest Rutherford. In 1930, Kapitza became director of the Mond Laboratory at Cambridge, which had been built for him. But when in 1934 he went to the USSR for a professional meeting, dictator Josef Stalin did not allow him to return. The Mond Laboratory was sold to the Soviet government at cost and transported to the Soviet Academy of Sciences for Kapitza's use. In 1946, he refused to work on the development of nuclear weapons and was put under house arrest until after Stalin's death in 1953. Kapitza was one of the first to study the unusual properties of helium II – the form of liquid helium that exists below 2.2 K ($-271.0°C/-455.7°F$). Helium II conducts heat far more rapidly than copper, which is the best conductor at ordinary temperatures, and Kapitza showed that this is because it has far less viscosity than any other liquid or gas. This property of helium is known as superfluidity. In 1939 Kapitza built apparatus for producing large quantities of liquid oxygen, used in steel production. He also invented a turbine for producing liquid air cheaply in large quantities.

Laue, Max Theodor Felix von (1879–1960)

German physicist who was awarded the Nobel Prize for Physics in 1914 for his pioneering work in measuring the wavelength of X-rays by their diffraction through the closely spaced atoms in a crystal. His work led to the techniques of X-ray spectroscopy, used in nuclear physics, and X-ray diffraction, used to elucidate the molecular structure of complex biological materials.

Russian physicist Peter Kapitza standing in front of the alternator he designed to produce strong magnetic fields of up to 72,000 amperes. AIP Emilio Segrè Visual Archives, Frenkel Collection

Laue was born near Koblenz and studied at Göttingen and Berlin. He was assistant to Max Planck at the Institute of Theoretical Physics in Berlin 1905–09, and worked at the Institute of Theoretical Physics in Munich 1909–14, when he became professor at Frankfurt. After World War I he became director of the Institute of Theoretical Physics in Berlin, resigning in 1943 in protest against Nazi policies. Although Laue had refused to participate in the German atomic-energy project, he was interned in Britain by the Allies after the war. He returned to Germany in 1946, and in 1951 became director of the Max Planck Institute for Research in Physical Chemistry. In 1912 Laue's idea of passing X-rays through crystals was first tested, and provided experimental demonstration of the nature of both crystal structure and X-ray radiation. He had initially considered only the interaction between the atoms in the crystal and the radiation, but later included a correction for the forces acting between the atoms.

Lawrence, Ernest O(rlando) (1901–1958)

US physicist. He was awarded the Nobel Prize for Physics in 1939 for his invention of the *cyclotron* particle accelerator which pioneered the production of artificial radioisotopes, the study of elementary particle interactions, and the synthesis of new transuranic elements.

During World War II Lawrence was involved with the separation of uranium-235 and plutonium for the development of the atomic bomb, and he organized the Los Alamos Scientific Laboratories at which much of the work on this project was carried out. After the war, he continued as a believer in nuclear weapons and advocated the acceleration of their development.

Lawrence was born in South Dakota and studied there and at Minnesota, Chicago, and Yale universities. He was professor of physics at the University of California, Berkeley, from 1930 and director from 1936 of the Radiation Laboratory, which he built into a major research centre for nuclear physics. The first cyclotrons were made in 1930 and were only a few centimetres in diameter. Each larger and improved design produced particles of higher energy than its predecessor, and a 68-cm/27-in model was used to produce artificial radioactivity. Among the results obtained from the use of the accelerated particles in nuclear transformations was the disintegration of the lithium nucleus to produce helium nuclei.

Lorentz, Hendrik Antoon (1853–1928)

Dutch physicist who helped to develop the theory of electromagnetism, which was recognized by the award (jointly with his pupil Pieter Zeeman) of the 1902 Nobel Prize for Physics.

Lorentz was born in Arnhem, the Netherlands, on 18 July 1853. He was educated at local schools and at the University of Leyden, which he left at the age of 19 to return to Arnhem as a teacher while writing his PhD thesis on the theory of light reflection and refraction. By the time he was 24 he was professor of theoretical physics at Leyden. He remained there for 39 years, before taking up the directorship of the Teyler Institute in Haarlem, where he was able to use the museum's laboratory facilities. He died in Haarlem on 4 February 1928.

Much of Lorentz's work was concerned with James Clerk Maxwell's theory of electromagnetism, and his development of it became fundamental to Albert Einstein's special theory of relativity. Lorentz attributed the generation of light by atoms to oscillations of charged particles (electrons) within them. This theory was confirmed in 1896 by the discovery of the Zeeman effect, in which a magnetic field splits spectral lines.

In 1904 Lorentz extended the work of George FitzGerald to account for the negative result of the Michelson–Morley experiment and produced the so-called Lorentz transformations, which mathematically predict the changes to mass, length, and time for an object travelling at near the speed of light.

McMillan, Edwin Mattison (1907–1991)

US physicist. In 1940 he discovered neptunium, the first transuranic element, by bombarding uranium with neutrons. He shared the Nobel Prize for Chemistry in 1951 with Glenn Seaborg for their discovery and work in the chemistry of transuranic elements.

In 1943 McMillan developed a method of overcoming the limitations of the cyclotron, the first accelerator, for which he shared, 20 years later, an Atoms for Peace award with I Veksler, director of the Soviet Joint Institute for Nuclear Research, who had come to the same discovery independently.

McMillan was born in Los Angeles and studied at the California Institute of Technology and Princeton. From 1932 he was on the staff at the University of California, as professor 1946–73, except during World War II when he worked on radar and Seaborg took up his work at Berkeley. The discovery of plutonium facilitated the construction of the first atom bomb, in which McMillan also took part.

Maiman, Theodore Harold (1927–)

US physicist who in 1960 constructed the first working laser.

Maiman was born in Los Angeles and studied at Columbia and Stanford universities. From 1955 to 1961 he worked at the Hughes Research Laboratories. In 1962 he founded the Korad Corporation to manufacture lasers; in 1968 he founded Maiman Associates, a laser and optics consultancy; he cofounded the Laser Video Corporation in 1972. In 1975 he joined the TRW Electronics Company, Los Angeles.

In 1955, Maiman began improving the maser (microwave amplifier), first designed in 1953 by US physicist Charles Townes. Townes had also demonstrated the theoretical possibility of constructing an optical maser, or laser, but Maiman was the first to build one. His laser consisted of a cylindrical, synthetic ruby crystal with parallel, mirror-coated ends, the coating at one end being semitransparent to allow the emission of the laser beam. A burst of intense white light stimulated the chromium atoms in the ruby to emit noncoherent red light. This red light was then reflected back and forth by the mirrored ends until eventually some of the light emerged as an intense beam of coherent red light – laser light. Maiman's apparatus produced pulses; the first continuous-beam laser was made in 1961 at the Bell Telephone Laboratories.

Meitner, Lise (1878–1968)

Austrian-born Swedish physicist who worked with the German radio-chemist Otto Hahn and was the first to realize that they had inadvertently achieved the fission of uranium. They also discovered protactinium in 1918. She refused to work on the atom bomb.

Meitner was born in Vienna and studied there and at Berlin. She joined Hahn to work with him on radioactivity at the Kaiser Wilhelm Institute for Chemistry, but was not given permission to work in the laboratory by their supervisor Emil Fischer, because she was a woman. Instead they had to set up a small laboratory in a carpenter's workroom. Nonetheless, Meitner was made joint director of the institute with Hahn in 1917 and was also appointed head of the Physics Department. In 1912, Meitner had also become an assistant to Max Planck at the Berlin Institute of Theoretical Physics, and she was made professor at Berlin in 1926. But in 1938 she was forced to leave Nazi Germany, and soon found a post at the Nobel Physical Institute in Stockholm. In 1947, a laboratory was established for her by the Swedish Atomic Energy Commission, and she later worked on an experimental nuclear reactor.

During the 1920s, Meitner studied the relationship between beta and gamma irradiation. She was the first to describe the emission of Auger electrons, which occurs when an electron rather than a photon is emitted after one electron drops from a higher to a lower electron shell in the atom.

In 1934 Meitner began to study the effects of neutron bombardment on uranium with Hahn. It was not found until after Meitner had fled from Germany that the neutron bombardment had produced not transuranic elements, as they expected, but three isotopes of barium. Meitner and her nephew Otto Frisch realized that the uranium nucleus had been split; they called it fission. A paper describing their analysis appeared in 1939.

Meitner continued to study the nature of fission products. Her later research concerned the production of new radioactive species using the cyclotron, and also the development of the shell model of the nucleus.

Millikan, Robert Andrews (1868–1953)

US physicist who was awarded the Nobel Prize for Physics in 1923 for his determination of Planck's constant (a fundamental unit of quantum theory) in 1916 and the electric charge on an electron in 1913.

His experiment to determine the electronic charge, which took five years to perfect, involved observing oil droplets, charged by external radiation, falling under gravity between two horizontal metal plates connected to a high-voltage supply. By varying the voltage, he was able to make the electrostatic field between the plates balance the gravitational field so that some droplets became stationary and floated. If a droplet of weight W is held stationary between plates separated by a distance d and carrying a potential difference V, the charge, e, on the drop is equal to Wd/V.

Millikan was born in Illinois and studied at Oberlin College and Columbia University, and in Germany with Max Planck at Berlin and Hermann Nernst at Göttingen. He worked at the University of Chicago 1896–21, becoming professor in 1910, and at the California Institute of Technology 1921–45.

Millikan also carried out research into cosmic rays, a term that he coined in 1925, when he proved that the rays do come from space.

Moseley, Henry Gwyn Jeffreys (1887–1915)

English physicist. From 1913 to 1914 he established the series of atomic numbers (reflecting the charges of the nuclei of different elements) that led to the revision of Russian chemist Dmitri Mendeleyev's periodic table of the elements.

English physicist Henry Moseley in Balliol Trinity Laboratory, Oxford, c. 1910. Moseley's fundamental discovery was a milestone in the knowledge of the constitution of the atom, and who knows what he might have discovered had he not been so tragically killed at so young an age. University of Oxford Museum of Science, courtesy AIP Emilio Segrè Visual Archives

Moseley was born in Weymouth, Dorset, and studied at Oxford. He worked in the Manchester laboratory of Ernest Rutherford, the pioneer of atomic science, 1910–13, and then at Oxford. Moseley was killed during the Gallipoli campaign of World War I. In 1913 Moseley introduced X-ray spectroscopy and found that the X-ray spectra of the elements were similar but the frequencies of corresponding lines changed regularly through the series of elements. A graph of the square root of the frequency of each radiation against the number representing the element's position in the periodic table gave a straight line. He called this number the *atomic number* of the element; the equation is known as *Moseley's law*. When the elements are arranged by atomic number instead of atomic mass, problems appearing in the Mendeleyev version are resolved. The numbering system also enabled Moseley to predict correctly that several more elements would be discovered.

Müller, K Alexander (1927–)

Swiss physicist who was awarded the Nobel Prize for Physics in 1987 for his work on high-temperature superconductivity in ceramic materials. The discovery of these materials was a significant step towards the use of superconductors in computers, magnetic levitation trains, and the more efficient generation and distribution of electricity. He shared the Nobel Prize with Georg Bednorz.

Superconductivity is the resistance-free flow of electrical current which occurs in many metals and metallic compounds at very low temperatures, within a few degrees of absolute zero (0 K/–273.16°C/–459.67°F). In 1986 Müller and Bednorz showed that a ceramic oxide of lanthanum, barium, and copper became superconducting at temperatures above 30 K, much hotter than for any previously known superconductor.

Müller was born in Basel, Switzerland, and studied physics and mathematics at the Swiss Federal Institute of Technology in Zürich. After graduating in 1958, he joined the Battelle Memorial Institute in Geneva. While in Geneva he was appointed lecturer (becoming a professor in 1970) at the University of Zürich. In 1963 he joined the IBM Zürich Research Laboratories at Rüschlikon.

Oppenheimer, J(ulius) Robert (1904–1967)

US physicist. As director of the Los Alamos Science Laboratory 1943–45, he was in charge of the development of the atom bomb (the Manhattan Project). He objected to the development of the hydrogen bomb, and was alleged to be a security risk in 1953 by the US Atomic Energy Commission (AEC).

Investigating the equations describing the energy states of the atom, Oppenheimer showed in 1930 that a positively charged particle with the mass of an electron could exist. This particle was detected in 1932 and called the positron.

Oppenheimer was born in New York and studied at Harvard, going on to postgraduate work with physicists Ernest Rutherford and J J Thomson at Cambridge, UK, and Max Born at Göttingen, Germany. Between 1929 and 1942 he was on the staff of both the University of California, Berkeley, and the California Institute of Technology. After World War II he returned briefly to California and then in 1947 was made director of the Institute of Advanced Study at Princeton University. Oppenheimer also served as chair of the General Advisory Committee to the AEC 1946–52. During World War II he reported to the Federal Bureau of Investigation friends and acquaintances who he thought might be communist agents; physicist David Bohm was one such.

Pauli, Wolfgang (1900–1958)

Austrian-born Swiss physicist who was awarded the Nobel Prize for Physics in 1945 for his discovery of the *exclusion principle*: in a given system no two fermions (electrons, protons, neutrons, or other elementary particles of half-integral spin) can be characterized by the same set of quantum numbers. He also predicted the existence of neutrinos.

The exclusion principle, announced in 1925, involved adding a fourth quantum number to the three already used (n, l, and m). This number, s, would represent the spin of the electron and would have two possible values. The principle also gave a means of determining the arrangement of electrons into shells around the nucleus, which explained the classification of elements into related groups by their atomic number.

Pauli was born in Vienna and studied in Germany at Munich. He then went to Göttingen as an assistant to German physicist Max Born, moving on to Copenhagen to study with Danish physicist Niels Bohr. From 1928 Pauli was professor of experimental physics at the Eidgenössische Technische Hochschule, Zürich, though he spent World War II in the USA at the Institute for Advanced Study, Princeton.

The neutrino was proposed in 1930 to explain the production of beta radiation in a continuous spectrum; it was eventually detected in 1956.

Peierls, Rudolf Ernst (1907–1995)

German-born British physicist who contributed to the early theory of the neutron–proton system. He helped to develop the atomic bomb 1940–46. He was knighted in 1968.

Peierls was born in Berlin and studied at universities in Germany and Switzerland under leading atomic physicists. From 1933 he worked in the UK, and was professor at Birmingham 1937–63 and at Oxford 1963–74.

In 1940, at Birmingham, Austrian physicist Otto Frisch and Peierls made an estimate of the energy released in a nuclear chain reaction, which indicated that a fission bomb would make a weapon of terrifying power. They drew the attention of the British government to this in 1940, and Peierls was placed in charge of a small group concerned with evaluating the chain reaction and its efficiency. In 1943, when Britain decided not to continue its work on nuclear energy, Peierls moved to the USA to help in the work of the Manhattan Project, first in New York and then at Los Alamos.

Planck, Max Karl Ernst Ludwig (1858–1947)

German physicist who was awarded the Nobel Prize for Physics in 1918 for his formulation of the quantum theory in 1900. His research into the manner in which heated bodies radiate energy led him to report that energy is emitted only in indivisible amounts, called 'quanta', the magnitudes of which are proportional to the frequency of the radiation. His discovery ran counter to classical physics and is held to have marked the commencement of the modern science.

Measurements of the frequency distribution of black-body radiation by Wilhelm Wien in 1893 showed the peak value of energy occurring at a higher frequency with greater temperature. This may be observed in the varying colour produced by a glowing object. At low temperatures, it glows red but as the temperature rises the peak energy is emitted at a greater frequency, and the colour become yellow and then white.

Wien attempted to derive a radiation law that would relate the energy to frequency and temperature but discovered a radiation law in 1896 that was valid only at high frequencies. Lord Rayleigh later found a similar equation that held for radiation emitted at low frequencies. Planck was able to combine these two radiation laws, arriving at a formula for the observed energy of the radiation at any given frequency and temperature. This entailed making the assumption that the energy consists of the sum of a finite number of discrete units of energy that he called quanta, and that the energy ε of each quantum is given by the equation: $\varepsilon = h\nu$, where ν is the frequency of the radiation and h is a constant now recognized to be a fundamental constant of nature, called Planck's constant. By directly relating the energy of a radiation to its frequency, an explanation was found for the observation that radiation of greater energy has a higher frequency distribution.

Planck's idea that energy must consist of indivisible particles, not waves, was revolutionary because it totally contravened the accepted belief that radiation consisted of waves. It soon found rapid acceptance: Albert Einstein in 1905 used Planck's quantum theory as an explanation for photo-electricity and in 1913 Danish physicist Niels Bohr successfully applied the quantum theory to the atom. This was later developed into a full system of quantum mechanics in the 1920s, when it also became clear that energy and matter have both a particle and a wave nature.

Planck's constant, a fundamental constant (symbol h), is the energy of one quantum of electromagnetic radiation divided by the frequency of its radiation.

Planck was born in Kiel and studied at Munich. He became professor at Kiel in 1885, but moved to Berlin in 1888 as director of the newly founded Institute for Theoretical Physics. He was also professor of physics at Berlin 1892–1926. Appointed president of the Kaiser Wilhelm Institute in 1930, he resigned in 1937 in protest at the Nazis' treatment of Jewish scientists. In 1945, after World War II, the institute was renamed the Max Planck Institute and moved to Göttingen. Planck was reappointed its president.

Röntgen, Wilhelm Konrad (1845–1923)

German physicist. He was awarded the Nobel Prize for Physics in 1901 for his discovery of X-rays in 1895. While investigating the passage of electricity through gases, he noticed the fluorescence of a barium platinocyanide screen. This radiation passed through some substances opaque to light, and affected photographic plates. Developments from this discovery revolutionized medical diagnosis.

Röntgen was born in Lennep, Prussia, and studied in Switzerland at the Zürich Polytechnic. He was professor at Giessen 1879–88, then director of the Physical Institute at Würzburg, and ended his career in the equivalent position at Munich 1900–20.

It was at Würzburg that Röntgen conducted the experiments that resulted in the discovery of X-rays (originally named after him). Today, the unit of radiation exposure is called the *roentgen*, or röntgen (symbol R). He refused to make any financial gain out of his findings, believing that the products of scientific research should be made freely available to all.

Röntgen worked on such diverse topics as elasticity, heat conduction in crystals, specific heat capacities of gases, and the rotation of plane-polarized light. In 1888 he made an important contribution to electricity when he confirmed that magnetic effects are produced by the motion of electrostatic charges.

Rutherford, Ernest (1871–1937)

1st Baron Rutherford of Nelson, New Zealand-born British physicist. He was a pioneer of modern atomic science. His main research was in the field of radioactivity, and he discovered alpha, beta, and gamma rays. He was the first to recognize the nuclear nature of the atom in 1911. He was awarded the Nobel Prize for Chemistry in 1908 for his work in atomic disintegration and the chemistry of radioactive substances.

Rutherford produced the first artificial transformation, changing one element to another (1919) by bombarding nitrogen with alpha particles and getting hydrogen and oxygen. After further research he announced that the nucleus of any atom must contain hydrogen nuclei; at Rutherford's suggestion, the name 'proton' was given to the hydrogen nucleus in 1920. He speculated that uncharged particles (neutrons) must also exist in the nucleus. In 1934, using heavy water, Rutherford and his co-workers bombarded deuterium with deuterons and produced tritium. This may be considered the first nuclear fusion reaction. He was knighted in 1914, and created Baron in 1931.

Rutherford was born near Nelson on South Island and studied at Christchurch. In 1895 he went to Britain and became the first research student to work under English physicist J J Thomson at the Cavendish Laboratory, Cambridge. In 1898 Rutherford obtained his first academic position with a professorship at McGill University, Montréal, Canada, which then boasted the best-equipped laboratory in the world. He returned to the UK in 1907, to Manchester University. From 1919 he was director of the Cavendish Laboratory, where he directed the construction of a particle accelerator, and was also professor of 'natural philosophy' at the Royal Institution from 1921. Rutherford began investigating radioactivity in 1897 and had by 1900 found three kinds of radioactivity with different penetrating power: alpha, beta, and gamma rays. When he moved to Montréal, he began to use thorium as a source of radioactivity instead of uranium. English chemist Frederick Soddy helped Rutherford identify its decay products, and in 1903 they were able to explain that radioactivity is caused by the breakdown of the atoms to produce a new element. In 1904 Rutherford worked out the series of transformations that radioactive elements undergo and showed that they end as lead. In 1914, Rutherford found that positive rays consist of hydrogen nuclei and that gamma rays are waves that lie beyond X-rays in the electromagnetic spectrum.

Salam, Abdus (1926–1996)

Pakistani physicist who was awarded the Nobel Prize for Physics in 1979 for his theory linking the electromagnetic and weak nuclear forces. He was the first person from his country to receive a Nobel Prize.

Salam shared the Nobel Prize with US physicists Sheldon Glashow and Steven Weinberg for unifying the theories of electromagnetism and the weak force, the force responsible for a neutron transforming into a proton, an electron, and a neutrino during radioactive decay. Building on Glashow's work, Salam and Weinberg separately arrived at the same theory in 1967.

Salam was born in Jhang near Faisalabad, in what was then part of British India. He attended Government College in Lahore before going to Cambridge University in the UK. From 1957 he was professor at Imperial College, London, and he was chief scientific adviser to the president of Pakistan 1961–74. Salam was also instrumental in setting up the International Centre for Theoretical Physics in Trieste, Italy, to stimulate science and technology in developing countries. The theory actually involves two new particles (the W^0 and B^0), which combine in different ways to form either the photon or the Z^0. This was verified experimentally at CERN, the European particle-physics laboratory near Geneva, in 1973, though the W and Z particles were not detected until 1983. Weinberg and Salam also predicted that the electroweak interaction should violate left–right symmetry and this was confirmed by experiments at Stanford University in California.

Pakistani theoretical physicist Abdus Salam whilst lecturing at Stanford University, California, in 1979. Abdus Salam became a scientist by accident, when he won a scholarship to Cambridge in 1945 from the Punjab Small Peasants' welfare fund; he had originally intended to join the Indian civil service. AIP Emilio Segrè Visual Archives, Segrè Collection

Schrieffer, John Robert (1931–)

US physicist who, with John Bardeen and Leon Cooper, was awarded the Nobel Prize for Physics in 1972 for developing the first satisfactory theory of superconductivity (the resistance-free flow of electrical current which occurs in many metals and metallic compounds at very low temperatures). He has also worked on ferromagnetism, surface physics, and dilute alloys.

In 1956 Cooper showed that, at low temperatures, electrons in a metal can weakly attract one another by distorting the lattice of metal atoms, forming a bound pair. In 1957 Bardeen, Cooper, and Schrieffer showed that these bound pairs of electrons can move through the metal without resistance. This theory of superconductivity (called the BCS theory) is amazingly complete and explains all known phenomena associated with superconductivitity in metals and alloys (except the high-temperature superconducting ceramics discovered in the late 1980s).

Schrieffer was born at Oak Park, Illinois, USA. He studied engineering and physics at the Massachusetts Institute of Technology (MIT), and the University of Illinois, where he worked under Bardeen for his doctorate (awarded in 1957) on superconductivity. Schrieffer became professor of physics at the University of Pennsylvania in 1964.

Schrödinger, Erwin (1887–1961)

Austrian physicist. He advanced the study of wave mechanics to describe the behaviour of electrons in atoms. In 1926 he produced a solid mathematical explanation of the quantum theory and the structure of the atom. He shared the Nobel Prize for Physics in 1933 for his work in the development of quantum mechanics.

Schrödinger's mathematical description of electron waves superseded matrix mechanics, developed in 1925 by Max Born and Werner Heisenberg, which also described the structure of the atom mathematically but, unlike wave mechanics, gave no picture of the atom. It was later shown that wave mechanics is equivalent to matrix mechanics.

Schrödinger was born and educated in Vienna. He was professor at Zürich, Switzerland, 1921–33. With the rise of the Nazis in Germany, he went to Oxford, UK, in 1933. Homesick, he returned to Austria in 1936 to take up a post at Graz, but the Nazi takeover of Austria in 1938 forced him into exile, and he worked at the Institute for Advanced Studies in Dublin, Ireland, 1939–56. He spent his last years at the University of Vienna. French physicist Louis de Broglie had in 1924, using ideas from Albert Einstein's special theory of relativity, shown that an electron or any other particle has a wave associated with it. In 1926 both Schrödinger and de Broglie published the same wave equation, which Schrödinger later formulated in terms of the energies of the electron and the field in which it was situated.

He solved the equation for the hydrogen atom and found that it fitted with energy levels proposed by Danish physicist Niels Bohr. In the hydrogen atom, the wave function describes where we can expect to find the electron. Although it is most likely to be close to one of the orbits of the Bohr atom, it does not follow a circular orbit but is described by the more complicated notion of an orbital, a region in space where the electron can be found with varying degrees of probability. Atoms other than hydrogen, and also molecules and ions, can be described by Schrödinger's wave equation, but such cases are very difficult to solve.

Schwinger, Julian Seymour (1918–1994)

US quantum physicist. His research concerned the behaviour of charged particles in electrical fields. This work, expressed entirely through mathematics, combines elements from quantum theory and relativity theory into a new theory called *quantum electrodynamics*, the most accurate physical theory of all time. Schwinger was awarded the Nobel Prize for Physics in 1965 for his development of the basic principles of quantum electrodynamics. He shared the award with Richard Feynman and Sin-Itiro Tomonaga.

Described as the 'physicist in knee pants', he entered college in New York at the age of 15, transferred to Columbia University and graduated at 17. At the age of 29 he became Harvard University's youngest full professor. He went to work on nuclear physics problems at Berkeley (in association with J Robert Oppenheimer) and at Purdue University. From 1943 to 1945 he worked on problems relating to radar at the Massachusetts Institute of Technology and, after the war, moved to Harvard, where he developed his version of quantum electrodynamics. He calculated the electron's anomalous magnetic moment respectively soon after its discovery. In 1957, Schwinger anticipated the existence of two different neutrinos associated with the electron and the muon, which was confirmed experimentally in 1963. He also speculated that weak nuclear forces are carried by massive, charged particles. This was confirmed in 1983 at CERN (the European Laboratory for Particle Physics) in Geneva. In 1972 Schwinger became professor of physics at the University of California, Los Angeles.

Segrè, Emilio Gino (1905–1989)

Italian-born US physicist who was awarded the Nobel Prize for Physics in 1959 for his discovery in 1955 of the antiproton, a new form of antimatter. He shared the award with his co-worker Owen Chamberlain. Segrè discovered the first synthetic element, technetium (atomic number 43), in 1937.

Segrè was born near Rome and studied there, working with Enrico Fermi. Segrè became professor at Palermo in 1936 but was forced into exile by the fascist government and, apart from wartime research at Los Alamos, New Mexico, worked from 1938 at the University of California at Berkeley, where he became professor in 1947.

In 1940 Segrè discovered another new element, now called astatine (atomic number 85). He again met up with Fermi, now at Columbia University, to discuss using plutonium-239 instead of uranium-235 in atomic bombs. Segrè began working on the production of plutonium at Berkeley and then moved to Los Alamos to study the spontaneous fission of uranium and plutonium isotopes.

In 1947, Segrè started work on proton–proton and proton–neutron interaction, using a cyclotron accelerator at Berkeley. This was how he detected the antiproton, which confirmed the relativistic quantum theory of English physicist Paul Dirac.

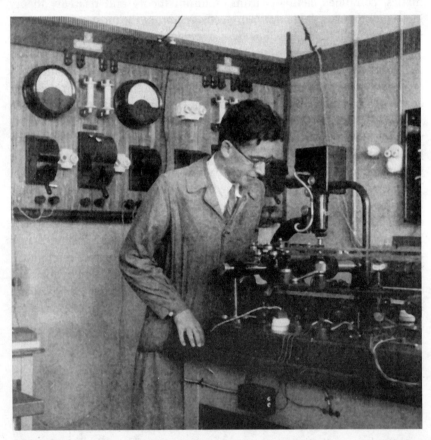

Italian-born US physicist Emilio Segrè working with a microphotometer in the laboratory of Dutch physicist Pieter Zeeman in Amsterdam in 1931. AIP Emilio Segrè Visual Archives, Segrè Collection

Shockley, William Bradford (1910–1989)

US physicist, computer scientist, and amateur geneticist who was awarded the Nobel Prize for Physics in 1956 for his study of semiconductors and the discovery of the transistor. He shared the award with his co-workers John Bardeen and Walter Brattain.

As well as research into transistor physics, Shockley's research has included theory on ferromagnetic domains and order and disorder in alloys.

Shockley was born in London, UK, to US parents who returned to the USA in 1913. He was educated at the California Institute of Technology and the Massachusetts Institute of Technology. In 1936 he joined Bell Telephone Laboratories, working in the group headed by US physicist Clinton Davisson, and remained there until 1955. He then became director of the Shockley Semiconductor Laboratory of Beckman Instruments, Inc, California, for research development and production of new transistor and other semiconductor devices. In 1963 he became first Alexander M Poniatoff professor of engineering science at Stanford University. At this time he began to pursue his interest in the origin of human intelligence, an area that embroiled him in much controversy owing to his recommendations for enforced sterilization of those with IQs below 100. During the 1970s Shockley was criticized for his claim that blacks were genetically inferior to whites in terms of intelligence.

Szilard, Leo (1898–1964)

Hungarian physicist and one of the 20th century's most original minds. He made major contributions to statistical mechanics, nuclear physics, nuclear engineering, molecular biology, political science, and genetics but was best known for his work on nuclear chain reaction. He was a central figure in the Manhattan Project and after the war became a strong proponent of the peaceful uses of atomic energy.

Leo Szilard was born in Budapest, the oldest of three children of a Jewish architect and engineer. A sickly child, he was mostly taught at home by his mother. He started studying electrical engineering in Budapest but his studies were interrupted when he was drafted into the Austro-Hungarian army in World War I. In 1920 he continued his studies in Berlin at the Technische Hochschule but transferred to read physics at the university, receiving a doctorate in 1922. His PhD thesis was on thermodynamics and the continuation of this work led to the publication of a famous 1929 paper establishing the connection between entropy and information. This was a forerunner to the theory of cybernetics.

In Berlin he was a research worker at the Kaiser Wilhelm Institute and a *Privatdozent* (unpaid lecturer) at the university. He worked on X-ray crystallography with Herman Marks and began work on patenting some of his pioneering inventions which included a prototype of the modern nuclear particle accelerator. With Albert Einstein he patented an electro-magnetic pump for liquid refrigerants, which is still used in nuclear reactors today.

In 1933, Szilard fled Hitler's Germany for the UK, working in the field of nuclear chain reactions at St Bartholomew's Hospital in London and the Clarendon Laboratory, Oxford. This work led to the establishment of the Szilard–Chambers reaction and the discovery of the emission of neutrons from beryllium. When Szilard emigrated to the USA in 1938, he learned that Otto Hahn and Lise Meitner had discovered fission in Germany. Recognizing the significance of this he persuaded Einstein (who commanded world respect) to write to President Roosevelt, warning him of the possibility of atomic bombs and encouraging him to develop them before the Germans did. This was the start of the Manhattan Project and initiated a programme that was to culminate in the dropping of a nuclear bomb on Hiroshima in Japan in 1945.

Szilard started work immediately at Columbia University to demonstrate the fission process and measure the number of neutrons released. In 1942 he joined Enrico Fermi at Chicago University, where they worked on the first controlled chain reaction on 2 December 1942. Fermi and Szilard were awarded the patent for the nuclear fission reactor in 1945. In the last months of the war, Szilard and others made a futile attempt to dissuade President Truman from dropping the bomb on Hiroshima, predicting the nuclear stalemate that would follow.

After the war Szilard moved into molecular biology, becoming professor of biophysics at Chicago in 1946. He was also prominent in the campaign for nuclear-arms control and in 1962 he founded the Council for a Livable World, a Washington lobby on arms control. He joined the Salk Institute of Biological Sciences at La Jolla in California in 1956, where he worked on genetics and immunology until his death on 30 May 1964.

Szilard was a fellow of the American Physical Society and the American Academy of Arts and Sciences and he received the Einstein Award in 1958 and the Atoms for Peace Award in 1959.

Thomson, J(oseph) J(ohn) (1856–1940)

English physicist. He discovered the electron in 1897. His work inaugurated the electrical theory of the atom, and his elucidation of positive rays and their application to an analysis of neon led to the discovery of isotopes. He was awarded the Nobel Prize for Physics in 1906 for his theoretical and experimental work on the conduction of electricity by gases. He was knighted in 1908.

Using magnetic and electric fields to deflect positive rays, Thomson found in 1912 that ions of neon gas are deflected by different amounts, indicating that they consist of a mixture of ions with different charge-to-mass ratios. English chemist Frederick Soddy had earlier proposed the existence of isotopes and Thomson proved this idea correct when he identified, also in 1912, the isotope neon-22. This work was continued by his student Francis Aston.

Thomson was born near Manchester and studied there and at Cambridge, where he spent his entire career. As professor of experimental physics 1884–1918, he developed the Cavendish Laboratory into the world's leading centre for subatomic physics. His son was George Paget Thomson.

Investigating cathode rays, Thomson proved that they were particulate and found their charge-to-mass ratio to be constant and with a value nearly 1,000 times smaller than that obtained for hydrogen ions in liquid electrolysis. He also measured the charge of the cathode-ray particles and found it to be the same in the gaseous discharge as in electrolysis. Thus he demonstrated that cathode rays are fundamental, negatively charged particles; the term 'electron' was introduced later.

Ting, Samuel Chao Chung (1936–)

US physicist. In 1974 he and his team at the Brookhaven National Laboratory, New York, detected a new subatomic particle, which he named the J particle. It was found to be identical to the ψ particle discovered in the same year by Burton Richter and his team at the Stanford Linear Accelerator Center, California. Ting and Richter shared the Nobel Prize for Physics in 1976 for their discovery of the ψ meson.

In 1996 Ting was at CERN and the Massachusetts Institute for Technology, working on the Alpha Magnetic Spectrometer, a spacebased antiparticle detector.

Walton, Ernest Thomas Sinton (1903–1995)

Irish physicist who collaborated with John Cockcroft on investigating the structure of the atom. They shared the Nobel Prize for Physics in 1951 for their success in 1932 in splitting the nucleus of an atom for the first time.

Walton and Cockcroft built the first successful particle accelerator. This used an arrangement of condensers to produce a beam of protons and was completed in 1932.

Walton was born in County Waterford and studied at Trinity College, Dublin, and 1927–34 at the Cavendish Laboratory in Cambridge, UK. He returned to Trinity and was professor there 1947–74.

Irish physicist Ernest Walton connects a filament in the first successful particle accelerator, which he constructed with English physicist John Cockcroft and completed in 1932.
British Information Services

Using the proton beam to bombard lithium, Walton and Cockcroft observed the production of large quantities of alpha particles, showing that the lithium nuclei had captured the protons and formed unstable beryllium nuclei which instantaneously decayed into two alpha particles travelling in opposite directions. They detected these alpha particles with a fluorescent screen. Later they investigated the transmutation of other light elements using proton beams, and also deuterons (nuclei of deuterium) derived from heavy water.

Weinberg, Steven (1933–)

US theoretical physicist best known for developing the unified electroweak theory. In 1957 he demonstrated with Pakistani theoretical physicist Abdus Salam that the weak nuclear force and the electromagnetic force (two of the fundamental forces of nature) are variations of a single underlying force that is now called the electroweak force. Weinberg, Salam, and US physicist Sheldon Glashow shared a Nobel Prize for Physics for this work in 1979.

Born in New York on 3 May 1933, the son of a court stenographer, Steven Weinberg was educated at the Bronx High School of Science in the same class as Glashow. With Glashow he entered Cornell in 1950, graduating in 1954 and then moving on to Princeton where he gained a PhD in 1957. He held positions at Columbia for two years, at Berkeley 1959–69, and at the Massachusetts Institute of Technology 1969–73. He was Higgins Professor of Physics at Harvard 1973–86 before becoming professor of physics and astronomy at the University of Texas at Austin.

At Berkeley in 1967, working with Glashow's early explanation, Weinberg produced a gauge theory that correctly predicted electromagnetic and weak nuclear forces. This was later to become known as the electroweak theory. In his paper, 'A model of leptons', he showed that although electromagnetism is much stronger than the weak force of everyday energies, the only way to devise a theory of the weak force is to include the electromagnetic force. Weinberg showed how what was seemingly impossible could be achieved and the forces could be unified through the interchange of particles in spite of the difference in their strengths. Abdus Salam had independently reached the same conclusions and what became known as the Weinberg–Salam model was a major advance on earlier models that had originally applied to leptons. The model's most striking prediction was that of the presence of neutral-current weak interactions (which had previously been believed to be absent). It also predicted the cross sections and other properties of a large number of neutral-current processes in terms of a single parameter, and together with the Glashow–Iliopoulis–Maiani charm scheme is the basis for an understanding of particle physics. The fundamental constituents of the model are six quarks and six leptons. The quarks interact through the strong and the electroweak interactions, but the leptons have only electroweak interactions. The strong interactions are mediated by a neutral boson, the gluon, and the theory that describes these interactions is called quantum chromodynamics, while the weak force is mediated by the W+ and W– boson. In 1979 Weinberg shared the Nobel Prize for Physics for this work with Salam and his old school friend Sheldon Glashow, who had extended the work that Weinberg and Salam had independently developed.

Weinberg also took an interest in cosmology and published *Gravitation and Cosmology* (1972). While he was at Harvard in 1979 he published *The First Three Minutes*, which described the early universe and was immensely popular with a general audience. In *Dreams of a Final Theory* (1993) he put the case that contemporary theories contained glimpses of an outline of a final theory. In the book he suggests that if the US government were to go ahead with the planned construction of the Superconducting Super Collider in Texas, it would be powerful enough to reveal the boson described by English physicist Peter Higgs (1929–) in 1964 and used by Weinberg to support his model.

Wigner, Eugene P(aul) (1902–1995)

Hungarian-born US physicist who shared the Nobel Prize for Physics in 1963 for his work, which introduced the notion of parity, or symmetry theory, into nuclear physics, showing that all nuclear processes should be indistinguishable from their mirror images.

While his earlier research concentrated on rates of chemical reactions, the theory of metallic cohesion, and group theory in quantum mechanics, he later focused on nuclear structure, including studies of nuclear resonance, electron spin, and the mirror nuclides, now known as *Wigner nuclides*. He also gave his name to several other atomic phenomena, such as the *Wigner effect*, a rapid rise in temperature in a nuclear reactor pile when, under particle bombardment, such materials as graphite deform, swell, then suddenly release large amounts of energy. This was the cause of the fire at the British Windscale plant in 1957.

Wigner was one of the scientists who persuaded President Roosevelt to commit the USA to developing the atom bomb. He took leave from Princeton University 1942–45, to join Enrico Fermi in Chicago, Illinois, where his calculations were essential to the design of the first atomic bomb, although he later said: 'Making a great weapon is not something to be proud of'. After World War II Wigner became an advocate of nuclear arms control and, in 1960, was awarded the Atoms for Peace Award in recognition of his vigorous support for the peaceful use of atomic energy.

Born in Budapest, Hungary, Wigner was educated at the city's Lutheran Gymnasium. In the 1920s he took up postgraduate studies in Berlin where he was present at Albert Einstein's seminars. In 1930 he emigrated to the USA (becoming a US citizen in 1937), and taught at Princeton University 1930–36 and the University of Wisconsin in 1937, before returning to Princeton as a professor of mathematics 1938–71. He won the 1963 Nobel Prize for his many contributions to nuclear physics and elementary particles; the prize was shared with Maria Goeppert-Mayer and Hans Jensen who worked separately. After retiring in 1971, he continued his interest in the effect of nuclear technology on society.

Yukawa, Hideki (1907–1981)

Japanese physicist. He was awarded the Nobel Prize for Physics in 1949 for his discovery in 1935 of the strong nuclear force that binds protons and neutrons together in the atomic nucleus, and his prediction of the existence of the subatomic particle called the meson.

Yukawa was born and educated in Kyoto and spent his career at Kyoto University, becoming professor in 1939 and director of the university's newly created Research Institute for Fundamental Physics from 1953. Yukawa's theory of nuclear forces postulated the existence of a nuclear 'exchange force' that counteracted the mutual repulsion of the protons and therefore held the nucleus together. He predicted that this exchange force would involve the transfer of a particle (the existence of which was then unknown), and calculated the range of the force and the mass of the hypothetical particle, which would be radioactive, with an extremely short half-life. The muon, discovered in 1936, was thought to be this particle, and was at first called the mu-meson. The pion, or pi-meson, discovered in 1947, was subsequently identified with Yukawa's particle. In 1936 Yukawa predicted that a nucleus could absorb one of the innermost orbiting electrons and that this would be equivalent to emitting a positron. These innermost electrons belong to the K electron shell, and this process of electron absorption by the nucleus is known as K capture.

Zweig, George (1937–   )

Russian-born US theoretical physicist who, along with US physicist Murray Gell-Mann, developed the quark theory. They developed a theory based on three quarks as the fundamental building blocks of the hadrons.

Zweig was born in Moscow but educated at the University of Michigan and California Institute of Technology (Caltech) from where he received his PhD in 1963. Zweig spent the next year at CERN before returning to Caltech, where he was made professor of physics in 1967. Zweig is currently professor of theoretical physics at CERN.

Part Two

Directory of Organizations and Institutions   103

Selected Works for Further Reading   121

Web Sites   127

Glossary   145

4 Directory of Organizations and Institutions

Abdus Salam Centre for Theoretical Physics (ICTP)

The ICTP was set up in 1964 by Abdus Salam, later a Nobel prizewinner, to provide facilities for theoretical physicists from Third World countries.

Address:
Strada Costiera 11
34100 Trieste, Italy
phone: +39 (0)40 2240111
fax: +39 (0)40 224163
e-mail: sci_info@ictp.trieste.it
Web site: http://www.ictp.trieste.it/

African Academy of Sciences

The African Academy of Sciences was conceived in 1985 when 22 prominent scientists met in Trieste, Italy, at the inauguration of the Third World Academy of Science. On 10 December 1995, its Constitution was ratified at a meeting held at the TWAS headquarters at the International Centre for Theoretical Physics (ICTP) in Trieste. Membership has grown rapidly from the initial 33 members in 1985 to 107 in 1995, representing 24 African countries and 5 countries overseas.

Address:
PO Box 14798
Nairobi, Kenya
phone: +254 2 884401
fax: +254 2 884406
e-mail: aas@arcc.permanet.org
Web site: http://www.sas.upenn.edu/African_Studies/Hornet/
menu_aas.html

American Association for the Advancement of Science (AAAS)

The AAAS is a nonprofit professional society dedicated to the advancement of scientific and technological excellence across all disciplines, and to the public's understanding of science and technology. AAAS is among the oldest societies in America, having been founded in Philadelphia in 1848. Anyone may join AAAS simply by paying membership dues. Many other scientific and engineering societies have chosen to become formally affiliated with AAAS. Currently there are 285 such affiliates, making AAAS the world's largest federation of scientific and engineering societies.

Address:
AAAS
1200 New York Avenue NW
Washington
DC 20005
USA
phone: +1 202 326 6400
e-mail: webmaster@aaas.org
Web site: http://www.aaas.org/

American Physical Society

The American Physical Society is an organization of more than 40,000 physicists worldwide, formed in 1899 and dedicated to the advancement and diffusion of the knowledge of physics. The APS publishes some of the world's leading physics research journals – the *Physical Review* series, Physical Review Letters, and Reviews of Modern Physics – and organizes scientific meetings where new results are reported and discussed.

Address:
APS Headquarters
One Physics Ellipse
College Park
MD 20740-3844, USA
phone: +1 301 209 3200
fax: +1 301 209 0865
Web site: http://www.aps.org/

Ames Research Center

Organization for the development of new aeronautics technologies. It was founded in 1939 as an aircraft research laboratory, and in 1958 became part of the National Aeronautics and Space Administration (NASA).

Address:
Moffett Field
CA 94035-1000, USA
phone: +1 650 604 5000
fax: +1 650 604 3445
Web site: http://www.arc.nasa.gov/

Bell Research Labs, AT&T

The most famous research centre run by a private company, Bell Labs has produced a long line of Nobel prizewinners, including William Shockley, John Bardeen, and Walter Brattain, who invented the transistor there in the late 1940s. Low-temperature physics is also an area in which outstanding research has been conducted here.

Web site: http://www.research.att.com/

Berkeley Lab

E O Lawrence, inventor of the cyclotron, founded this, the oldest of the US national laboratories, in 1931. Although best known as a mecca of particle physics, Berkeley Lab has a broader sweep. Research is conducted in advanced materials, life sciences, energy efficiency, particle detectors, and particle accelerators. The lab is managed by the University of California for the US Department of Energy.

Address:
Lawrence Berkeley National Lab
1 Cyclotron Road
Berkeley
CA 94720
USA
phone: +1 510 486 4387
fax: +1 510 486 7000
Web site: http://www.lbl.gov/

Brookhaven National Laboratory

The Brookhaven particle accelerator lab was the site of the first proton synchrotron, dubbed the Cosmotron because it accelerated particles to energies of 3 GeV, previously observed only in cosmic rays.

Address:
Long Island
NY 11973-5000
USA
phone: +1 631 344 8000
Web site: http://www.bnl.gov

California Institute of Technology

Located in Pasadena, California, Caltech is a small university dedicated to exceptional instruction and research in engineering and science. The student body is composed of 900 undergraduate and 1,100 graduate students. The outstanding faculty includes several Nobel laureates and off-campus facilities include the Jet Propulsion Laboratory, Palomar Observatory, and the Keck Observatory.

Address:
1200 East California Boulevard
Pasadena
CA 91125
USA
phone: +1 626 395 6811
Web site: http://www.caltech.edu/

Cavendish Laboratory

Established as a home for experimental physics in Cambridge University. The golden age of the Cavendish was the period from 1919 to 1937 when Ernest Rutherford was its director, and he and its students revealed the structure of the atom. Its current research effort is concentrated in three main fields: astrophysics and cosmology; high-energy physics; condensed matter physics.

Address:
Department of Physics
Cavendish Laboratory
Madingley Road
Cambridge
CB3 0HE
UK
phone: +44 (0)1223 337200
fax: +44 (0)1223 363263
e-mail: www@phy.cam.ac.uk
Web site: http://www.phy.cam.ac.uk

CERN

The European Laboratory for Particle Physics – originally Centre Européenne pour la Recherche Nucléaire (European Centre for Nuclear Research); it is always known by these initials. Its huge accelerator rings straddle the French–Swiss border.

Address:
CERN
Press Office
CH-1211 Geneva 23
Switzerland
phone : +41 22 76 76111
fax: +41 22 76 76555
Web site: http://www.cern.ch/CERN/

Cornell University

The high-energy physics laboratory at Cornell is the home of CESR, the Cornell Electron Storage Ring, which can store electrons at energies of up to 8 GeV, high enough to study the properties of bottom quarks.

Address:
Cornell University
Information and Referral Center
Day Hall Lobby, Ithaca
NY 14853-2801, USA
phone: +1 607 254 4636
Web site: http://www.cornell.edu/

Daresbury Laboratory

Part of the Central Laboratory of the Research Councils (CCLRC), one of Europe's largest multidisciplinary research support organizations. Daresbury has several major research facilities, and is a key centre in the European effort in controlled fusion, which promises to tame the power of the H-bomb for peaceful uses.

Address:
Daresbury Laboratory
Daresbury
Warrington
Cheshire WA4 4AD, UK
phone: +44 (0)1925 603000
fax +44 (0)1925 603100
e-mail: enquiries@circ.ac.uk
Web site: http://www.clrc.ac.uk/Who/PPR

Deutsches Elektronen-Synchrotron (DESY)

The laboratory performs basic research in high-energy and particle physics as well as in the production and application of synchrotron radiation. The laboratory has locations in Hamburg (this server) and Zeuthen, Germany, and is a member of the association of national research centres Hermann von Helmholtz-Gemeinschaft Deutscher Forschungszentren (HGF).

Address:
Notkestrasse 85
22607 Hamburg
Germany
phone: +49 40 8998 3613
fax: +49 40 8998 4307
e-mail: desypr@desy.de
Web site: http://www.desy.de/

European Physical Society

The European Physical Society, founded in 1968, is an organization of more than 70,000 European physicists. It provides a forum for individual physicists, and acts as a federation of national societies and academies. It also brings together university and industrial physicists working in emerging new technologies.

Address:
European Physical Society Secretariat Secretary General: David Lee
Rue Marc Seguin 34 F – 68060 Mulhouse Cedex
France
phone: + 33 389 329440
fax: + 33 389 329449
Web site: http://epswww.epfl.ch

Fermi National Accelerator Laboratory (Fermilab)

Fermilab is a US Department of Energy national laboratory for research exploring the fundamental nature of matter and energy. It contains the 6.4-km/4-mi circle of the Tevatron, the world's most powerful particle accelerator, and the smaller oval of the Main Injector, a new accelerator that began operating in 1999. Universities Research Association, Inc operates Fermilab under contract with the United States Department of Energy.

Address:
Fermilab
PO Box 500
Batavia
IL 60510-0500
USA
phone: +1 630 840 3000
fax: +1 630 840 4343
Web site: http://www.fnal.gov/

Harvard–Smithsonian Center for Astrophysics

The Harvard–Smithsonian Center for Astrophysics (CfA) is a joint collaboration between the Smithsonian Astrophysical Observatory and the Harvard College Observatory. The CfA's research mission is the study of the origin, evolution, and ultimate fate of the universe. More than 300 professional scientists, supported by some 500 technical, engineering, and administrative staff, pursue a broad range of research. The Harvard University Astronomy Department is also housed at the CfA. Divisions and Departments include: atomic and molecular physics; high energy astrophysics; optical and infrared astronomy; planetary sciences; radio and geoastronomy; solar and stellar physics; and theoretical astrophysics.

Address:
60 Garden Street
Cambridge
MA 02138
USA
Web site: http://cfa-www.harvard.edu/sao-home.html

Institut de Génie Atomique (IGA)

The Institute of Atomic Engineering in the Physics Department of the University of Lausanne is a leader in the field of condensed-matter research. It comprises four laboratories: the Nuclear Reactor Physics Laboratory, the Semicrystalline Solids Physics Laboratory, Metallurgical Physics Laboratory 1 (nanomechanics, nonlinear dynamics and superconductivity, internal friction), and Metallurgical Physics Laboratory 2 (shape memory alloys and ion implantation, plasticity and dislocations, cavitation, and erosion).

Address:
Swiss Federal Institute of Technology – Lausanne
Département de Physique
PHB Ecublens
CH-1015 Lausanne
Switzerland
phone: +41 21 693 3374/5
fax: +41 21 693 4444
Web site: http://igahpse.epfl.ch/iga2_english.html

Institute for Advanced Study

The Institute for Advanced Study is an independent private institution founded in 1930 by Louis Bamberger and Caroline Bamberger Fuld as a centre where intellectual inquiry can be carried out in the most favourable circumstances. The Institute has been home to some of the most highly regarded thinkers of the 20th century. It consists of the School of Historical Studies, the School of Mathematics, the School of Natural Sciences, and the School of Social Science. More than a dozen Nobel laureates have been Institute or faculty members, and many more are winners of the Wolf or MacArthur prizes, or the Fields Medal. The Institute has no formal curriculum, degree programmes, schedule of courses, laboratories, or other experimental facilities. It is committed to exploring the most fundamental areas of knowledge, areas where there is little expectation of immediate outcomes or striking applications.

Address:
Institute for Advanced Study
Einstein Drive
Princeton
NJ 08540-0631, USA
phone: +1 609 734 8000
fax: +1 609 924 8399
Web site: http://www.ias.edu/

Institute for Plasma Research

The Institute was established near Ahmedabad, in western India, in 1986 to carry out experimental and theoretical research in plasma physics, with emphasis on the physics of magnetically confined hot plasmas and nonlinear plasma phenomena. These activities are aimed at controlling nuclear fusion, a major research objective of the Indian government.

Address:
Institute For Plasma Research
Bhat
Gandhinagar 382428
Gujurat
India
phone: +91 079 9001/2/3
fax: +91 079 286 9017
Web site: http://www.plasma.ernet.in/

Institute of Fluid Mechanics, Göttingen

The Institute of Fluid Mechanics of the University of Göttingen, Germany, is a leader in the field of theoretical and applied aerodynamics. It conducts research in numerous areas, including industrial aerodynamics, high-speed aerodynamics, turbulence, turbine aerodynamics, scientific visualization, and expert computer systems for aerodynamics.

Address:
DLR Göttingen
Abt. SM-SK-HGA
Bunsenstrasse 10
37073 Göttingen
Germany
phone: +49 551 709 2178
fax: +49 551 709 2889
Web site: http://www.sm.go.dlr.de/

Institute of Physics

A learned society and the professional body for physicists in Great Britain and Ireland. With over 22,000 members, it looks after the interests of professional physicists and supports the education of the physicists of tomorrow. Institute of Physics Publishing, a wholly-owned subsidiary, is a major publisher of physics.

Address:
Institute of Physics
76 Portland Place
London
W1B 1NT
UK
phone: +44 (0)20 7470 4800
fax: +44 (0)20 7470 4848
e-mail: physics@iop.org
Web site: http://www.ioppublishing.com/IOP/

International Association of Mathematical Physics

The Association was founded in 1976 in order to promote research in mathematical physics. Mathematicians and physicists (including students) may become members. The Association sponsors a Congress every three years. Its members have interests ranging over diverse topics including quantum field theory, statistical physics, relations between geometry and physics, disordered systems, non-equilibrium systems, relativity and gravitation, dynamical systems and fluid dynamics. Unsolved problems are shared on the Association's Web site.

Web site: http://www.iamp.org/index.html

International Bureau of Weights and Measures

Body set up in Paris in 1875 (originally, Bureau International des Poids et Mesures) that still coordinates world standards for mass, length, time, electric current, and other quantities. It created the International System of Units (SI units) now used universally by scientists.

Address:
Pavillon de Breteuil
92312 Sèvres Cedex
France
phone: +33 1 4507 7070
fax: +33 1 4534 2021
e-mail: info@bipm.fr
Web site: http://www.bipm.fr/enus/

Jet Propulsion Laboratory

Organization managed by the California Institute of Technology (Caltech) for NASA. It is the primary US centre for robotic exploration of the Solar System. JPL spacecraft have visited all known planets with the exception of Pluto.

Address:
4800 Oak Grove Drive
Pasadena
CA 91109
USA
phone: +1 818 354 4321
Web site: http://www.jpl.nasa.gov

Joint Institute For Nuclear Research (JINR)

JINR, not far from Moscow, was established in March 1956 in order to study the fundamental properties of matter. It is now a large international scientific centre with seven laboratories. The main fields of the institute's research are: theoretical physics; elementary particle physics; nuclear physics; heavy ion physics; condensed matter physics; radiation biology and nuclear medicine; experimental instruments and methods. The institute's experimental facilities include pulsed neutron sources and particle accelerators. JINR specialists collaborate in experiments performed at CERN and US accelerators.

Address:
JINR
Joliot-Curie 6
141980 Dubna
Moscow region
Russian Federation
phone: +7 09621 65 059
fax: +7 09621 65 891
e-mail: post@office.jinr.dubna.su
Web site: http://www.jinr.dubna.su/

Kharkov Institute of Physics and Technology (KIPT)

Formerly called the Ukrainian Institute of Physics and Technology, KIPT possesses many unique experimental facilities, among which are a number of particle accelerators, and thermonuclear installations. Research work at KIPT includes: solid-state physics; physics of radiation effects; materials science; plasma physics and controlled fusion; nuclear physics; physics and engineering of electron accelerators; theoretical physics.

Address:
NSC KIPT
1 Akademicheskaya St
Kharkov, 361108, Ukraine
phone: +380 57 235 3530
fax: +380 57 235 1688
Web site: http://www.kipt.kharkov.ua/

Laboratory Of Molecular Biophysics

Part of the University of Oxford Department of Biochemistry that studies complex biological macromolecules and macromolecular assemblies. The major research techniques used are protein crystallography and molecular dynamics simulations. These methods are complemented by other physical and biochemical techniques: gene cloning, protein expression and purification, protein crystallization, analysis of protein-drug interactions, molecular modelling, and electron microscopy. Target biological systems include cell cycle proteins, viruses, and viral proteins.

Address:
Laboratory of Molecular Biophysics
University of Oxford
Rex Richards Building
South Parks Road
Oxford OX1 3QU, UK
phone: +44 (0)1865 275366
fax: +44 (0)1865 510454
Web site: http://biop.ox.ac.uk

Laboratory of Molecular Biology

In 1947 the Medical Research Council set up a unit for research on the molecular structure of biological systems to develop the use of X-ray diffraction to study proteins. Its first great success came in 1953 when James Watson and Francis Crick used data from Rosalind Franklin's X-ray diffraction work to propose the double-helix structure of DNA. Since then the laboratory has been a prolific source of new discoveries, in which physical techniques such as X-ray crystallography and electron micrography play a key role.

Address:
Laboratory of Molecular Biology
Medical Research Council Centre
Hills Road, Cambridge CB2 2QH, UK
phone: +44 (0)1223 248011
fax: +44 (0)1223 213556
Web site: http://www2.mcr-lmb.cam.ac.uk

Lawrence Livermore National Laboratory

Lawrence Livermore was established in 1952, under the management of the University of California, for the design, development, and stewardship of nuclear weapons. National security continues to be the Laboratory's defining responsibility. However, the Laboratory has additional programmes in energy, the environment, and bioscience to improve human health. The technical expertise acquired in defence work has also been used in non-military technology ranging from uranium enrichment to space research. The current budget is $1.3 billion, and the laboratory has 7,400 employees.

Address:
7000 East Ave
Livermore
CA 94550-9234
USA
phone: +1 925 423 3125
Web site: http://www.llnl.gov/

Linear Accelerator Center (SLAC)

Established in 1962, SLAC is a world research facility with 1,200 employees on site and 2,800 visiting researchers from around the world. Stanford University operates SLAC on behalf of the US Department of Energy. SLAC has launched, with the Lawrence Berkeley and Lawrence Livermore National Laboratories, an advanced research facility called the B Factory. Here electrons and positrons are collided to produce ephemeral particles called B mesons, to study, among other questions, the reason for the matter–antimatter imbalance in the universe.

Address:
Stanford University
Stanford
CA 94305
USA
phone: +1 650 723 2300
Web site: http://www.stanford.edu

Los Alamos National Laboratory

Los Alamos was first set up as the home of the Manhattan Project to build the first nuclear bomb. It later became a major centre for research in both military and civilian uses of nuclear energy. It is operated by the University of California for the US Department of Energy.

Address:
PO Box 1663
Los Alamos
NM 87545
USA
phone: +1 505 667 5061
e-mail: www-core@lanl.gov
Web site: http://www.lanl.gov/worldview/

Massachusetts Institute of Technology

This outstanding research and teaching establishment carries out studies in: astrophysics; atomic, molecular, and optical physics; biological and medical physics; condensed matter physics; elementary particle physics; nuclear physics; plasma physics; and theoretical physics. It has established a number of interdisciplinary laboratories, including the Research Laboratory of Electronics, the Laboratory of Nuclear Science, the Center for Space Research, the Center for Materials Science and Engineering, the Magnet Lab, and the Plasma Fusion Center.

Address:
77 Massachusetts Avenue
Room 6–113
Cambridge
MA 01239-4301, USA
phone: +1 617 253 4801
fax: +1 617 253 8554
Web site: http://web.mit.edu/physics/

Max Planck Society for the Advancement of Science

The Max Planck Society is a non-governmental, non-profit research organization. The Administrative Headquarters are located in Munich. It embraces 80 institutes, research centres, laboratories and project groups. These conduct basic research in the sciences, arts, and humanities, concentrating in particular areas of research to supplement research carried out by the universities. It makes numerous awards for outstanding research.

Address:
Max-Planck-Gesellschaft zur Förderung der Wissenschaften
eV Generalverwaltung Postfach 10 10 62
D-80084 München
Germany
phone: +49 89 2108 0
fax: +49 89 2108 1111
Web site: http://www.mpg.de/english/kontakt/

Medical Research Council Laboratory, Cambridge

Leader in the X-ray crystallography of large biological molecules; the structure of DNA was unravelled here.

Address:
Laboratory of Molecular Biology
Medical Research Council Centre
Hills Road
Cambridge
CB2 2QH, UK
phone: +44 (0)1223 248011
fax: +44 (0)1223 213556
Web site: http://www2.mrc-lmb.cam.ac.uk/

National Academy of Sciences

For advice on scientific issues relevant to policy decisions, US leaders often turn to the institution that was specially created for this purpose: the National Academy of Sciences, founded in 1863, and its sister organizations – the National Academy of Engineering, the Institute of Medicine, and the National Research Council. These non-profit organizations provide a public service by working outside the framework of government to ensure independent advice on matters of science, technology, and medicine. Hundreds of policy studies are produced each year. These reports examine a range of issues, from AIDS to obesity to science education, nuclear waste, and more.

Address:
2101 Constitution Avenue NW
Washington
DC 20418, USA
Web site: http://www.nas.edu

National Institution for Standards and Technology (NIST)

Formerly the National Bureau of Standards, this laboratory is concerned with laying down accurate standards for quantities including electrical current and resistance, time, and temperature, for use by organizations in US industry and science.

Address:
100 Bureau Drive
Stop 3460
Gaithersburg
MD 20899-3460, USA
phone: +1 301 975 6478
Web site: http://www.nist.gov/

National Physical Laboratory (NPL)

NPL is the UK's national measurement standards laboratory. It holds and maintains reference standards for the basic units of mass, length, time, temperature, luminous intensity, and electrical current, as well as many of the derived units. It also undertakes research and development to meet the needs of new industries, and provides consultancy, advice, and technology transfer to support commercial integrity and competitiveness.

Address:
Teddington
Middlesex
TW11 0LW, UK
phone: +44 (0)20 8943 6880
fax: +44 (0)20 8943 6458
e-mail: enquiry@npl.co.uk
Web site: http://www.npl.co.uk/npl/about/brief.html

National Science Foundation (NSF)

The National Science Foundation is an independent US government agency responsible for promoting science and engineering through programmes that invest over $3.3 billion per year in almost 20,000 research and education projects in science and engineering. The agency operates no laboratories itself but does support National Research Centers, certain oceanographic vessels, and Antarctic research stations. The Foundation also supports cooperative research between universities and industry and US participation in international scientific efforts.

Address:
4201 Wilson Boulevard
Arlington, VA 22230, USA
phone: +1 703 306 1234
Web site: http://www.nsf.gov/

National Society of Black Physicists (NSBP)

NSBP is the largest and most recognizable organization devoted to the African-American physics community. The purpose of NSBP is to promote the professional well-being of African-American physicists within the scientific community and within society at large. It also seeks to develop activities and programmes that highlight and enhance the benefits of the contributions that African-American physicists provide for the world community. Regular Membership in NSBP is open to those physicists from all ethnic groups who affirm the goals of the organization. Nonphysicists who support the goals of the organization also are welcomed as associate members.

Address:
Dr Sekazi Mtingwa
NSBP Office Manager
Department of Physics
North Carolina A&T State University
Greensboro
NC 27411-1086
USA
phone: +1 910 334 7646
fax: +1 910 334 7283
Web site: http://www.nsbp.org/history.html

Niels Bohr Institute

The Niels Bohr Institute was founded on 3 March 1921 by the great pioneer of quantum physics. The Institute's research activities are: theoretical nuclear physics; theoretical high-energy physics; condensed matter physics; physics of complex systems; biological physics; astrophysics; computing; experimental nuclear physics; experimental high-energy physics; and Mössbauer studies of meteorites.

Address:
Niels Bohr Institute Blegdamsvej
17 DK-2100 Copenhagen
Denmark
phone: +45 35 32 52 09
fax: +45 35 32 50 16
Web site: http://www.nbi.dk/

Novosibirsk State University Physics Department

The department collaborates with the Siberian Branch of the Russian Academy of Science, and houses institutes specializing in: nuclear physics, semiconductor physics, chemical kinetics and combustion, hydrodynamics, theoretical and applied mechanics, thermophysics, inorganic chemistry, and laser physics.

Address:
630090 Novosibirsk-90
Pirogova Str 2
PF, NSU
Russian Federation
phone: +7 3832 353854/356233
fax: +7 383 352653
e-mail: ff@ff.nsu.nsk.su
Web site: http://www.nsu.ru/english/

Royal Society of London

The Royal Society, founded in 1660, is an independent learned society, self-governing under a Royal charter, for the promotion of the natural sciences, including mathematics and the scientific aspects of engineering, agriculture, and medicine. Each year, up to 40 new fellows and six new foreign members are elected from amongst the most distinguished scientists in the British Commonwealth and other countries. The society encourages scientific endeavour by providing research appointments in British universities and grants for national and international scientific activities, hosts meetings and lectures, promotes the public understanding of science, and provides independent scientific advice. It has exchange agreements with scientific bodies in some 50 countries.

Address:
6 Carlton House Terrace
London
SW1Y 5AG, UK
phone: +44 (0)20 7839 5561 ext 2584 or 2507
fax: +44 (0)20 7930 2170/+44 (0)20 7451 2692
e-mail: info@ulcc.ac.uk
Web site: http://www.royalsoc.ac.uk

Rutherford Appleton Laboratory (RAL)

The UK's Rutherford Appleton Laboratory (RAL) was founded in 1957 and is now part of the Central Laboratory of the Research Councils. Its research programme includes astronomy, biology, chemistry, computing, work on new energy sources, engineering, environmental research, materials science, particle physics, astrophysics, and other areas of physics, radio communications, and space science. Facilities include ISIS, the world's most powerful pulsed neutron source, high-power lasers, high-performance computing and the Central Microstructure Facility.

Address:
Chilton, Didcot
Oxfordshire
OX11 0QX, UK
phone: +44 (0)1235 821900
fax: +44 (0)1235 445808
Web site: http://hepwww.rl.ac.uk/

TRIUMF

Canada's National Meson Research Facility is operated as a joint venture by Simon Fraser University and the universities of Alberta, Victoria, and British Columbia, with several associated universities and the support of the National Research Council of Canada. The giant TRIUMF cyclotron, the world's largest, generates intense beams of pions. TRIUMF is located on the University of British Columbia campus and provides world-leading facilities for experiments in subatomic research with beams of pions, muons, protons, and neutrons. Research areas include muon-catalysed hydrogen fusion, superconductor research, contraband detection systems, new radioisotopes for medical diagnosis, positron emission tomography (PET), proton therapy for tumours, smokestack emission control, and superfast microchips.

Address:
4004 Wesbrook Mall
Vancouver
BC V6T 2A3, Canada
phone: +1 604 222 1047
voice mail: +1 604 222 7347
fax: +1 604 222 1074
Web site: http://www.triumf.ca/homepage.html

University of Cambridge Institute of Astronomy
The IoA is a department of the University of Cambridge and is engaged in teaching and research in theoretical and observational astronomy. Theoretical problems studied range from models of quasars and of the evolution of the universe to theories of the formation and evolution of galaxies and stars, X-ray sources, and black holes. Much observational work centres around the use by staff of large telescopes abroad and in space to study quasars, galaxies, and the chemical constitution of stars. Instrumentation development is also an important area of activity, involving charge coupled devices and detector arrays for rapid recording of very faint light, and the design and construction of novel spectrographs.

Address:
Institute of Astronomy
University of Cambridge
Madingley Road
Cambridge
CB3 0HA, UK
phone: +44 (0)1223 337548
fax: +44 (0)1223 337523
Web site: http://www.ast.cam.ac.uk

5   Selected Works for Further Reading

Bernstein, Jeremy *Einstein,* 1973

For the non-specialist who wants an insight into the complex character rather than his complex theories.

Brown, Andrew *The Neutron and the Bomb: A Biography of Sir James Chadwick,* 1997

A thorough and sympathetic account.

Calder, Nigel *Einstein's Universe,* 1979

A full account of the history of relativity theory and its implications for the way we understand the universe. Included is enough material about Einstein for the reader to get an idea of what the man was like.

Carrigan, Richard and Trower, Peter (eds) *From Quarks to the Cosmos: Tools of Discovery,* 1990

A collection of articles from *Scientific American,* mainly from the 1980s, which brings particle physics into the 1990s. All are good, some are classics.

Close, Frank *The Particle Explosion,* 1983

A first guide to particle physics by a gifted and inspirational lecturer who presented the Royal Institution's Christmas lectures 1993–94. The many illustrations include photographs, diagrams, and the author's own cartoon quarks.

Close, Frank, Marten, Michael, and Sutton, Christine *The Forces of Nature,* 1987

A highly illustrated, loosely historical account describing the different particles and how they were discovered. At the same time it reveals how the field has developed with advances in experimental techniques for producing particles and tracking them down.

Coleman, James A *Relativity for the Layman,* 1969

One of the great classics of popular science. Written with a sure economy of words, this is a book of great clarity and elegance.

Cotterill, R *The Cambridge Guide to the Material World*, 1985

An extraordinarily wide and copiously illustrated survey of the varieties of matter from fundamental particles to the constituents of plants and animals, taking in crystals, liquids, glasses, polymers, and many others.

Davies, Paul *Particles and Forces: At the heart of the matter*, 1986

A valuable companion to any of the books that are mainly about quarks, that is, 'building blocks' of matter, this concentrates on the 'mortar'. It provides an introduction to how modern physics deals with forces at a quantum level.

Davies, Paul and Brown, Julian (eds) *Superstring: A theory of everything?*, 1988

A collection of discussions with leading physicists on the topic of super-string theory. Whilst the topic is at the cutting edge of theoretical physics it is presented is such a way as to be accessible to the general reader but challenging enough for the student of physics.

Einstein, Albert and Infeld, L *The Evolution of Physics*, 1938

The greatest of modern physicists describes the developments that he pioneered – relativity and quantum physics – as the outcome of earlier achievements.

Feynman, Richard *QED: The strange theory of light and matter*, 1985

QED (quantum electrodynamics) is a very strange theory that explains how light and electrons interact. The reader is treated to Feynman at his best, talking about his beloved theory and making it accessible to all.

Feynman, Richard *The Character of Physical Law*, 1965

Based on a television series by a great physicist and brilliant expositor. Concentrates on basic concepts such as conservation, symmetry, the concept of time, and probability in quantum physics.

Fritsch, Harald *Quarks, the Stuff of Matter*, 1984

A straightforward account of the development of quark theory, by one of those involved at the cutting edge.

Gleick, James *Genius*, 1992

Full-scale biography of the attractive and brilliant Richard Feynman.

Goldstein, M and Goldstein I F *The Refrigerator and the Universe*, 1993

Energy and entropy in physics, chemistry, and cosmology (including the greenhouse effect).

Gribbin, John *Schrödinger's Kittens*, 1995

The prolific science writer takes a look at the behaviour of light and searches for quantum and relativistic interpretations.

Harbison, James P and Nahory, Robert E *Lasers: Harnessing the atom's light*, 1998

An elegant book that covers history and workings of the laser, as well as current research, accessible for nonspecialists with some grounding in science.

Hawking, Stephen W *A Brief History of Time*, 1988

Best-selling account of the universe by a brilliant theoretical physicist with a talent for sharing his subject.

Hey, Tony and Walters, Patrick *Einstein's Mirror*, 1997

An ideal text for the newcomer to relativity to dip into. The large, almost coffee table, format leads the reader very gently and avoids the need for mathematics.

Johnson, George *Strange Beauty*, 2000

Excellent biography of the physicist Murray Gell-Mann.

Kaku, Michio *Hyperspace*, 1994

An introduction to some of the most exciting areas of modern physics, including black holes, multiple universes, and ten dimensional space. It requires no previous knowledge and no mathematics.

Kane, Gordon *Q is for Quantum: The A–Z of particle physics*, 1994

A fascinating up-to-date introduction to particle physics, which also presents a personal view of the field as it headed towards the 21st century. It places the present state of understanding in better context than many other books and articles.

Lederman, Leon and Schramm, David *Dreams of a Final Theory*, 1989

An experimental particle physicist and a theoretical astrophysicist team up to show how the physics of the very small has become inextricably linked with our understanding of the universe on cosmic scales. Well illustrated, the emphasis is on what we know through experiment.

Mendelssohn, K *The Quest for Absolute Zero: Meaning of low temperature physics*, 1977

A non-mathematical account of how very low temperatures are produced, and the phenomena such as superconductivity and super-fluidity that occur at these extreme conditions.

Molinari, A and Ricci, R A (eds) *From Nuclei to Stars: A meeting in nuclear physics and astrophysics exploring the path opened by H A Bethe*, 1986

Proceedings of the International School of Physics 'Enrico Fermi' course 91, June, 1984.

Rhodes, R *The Making of the Atomic Bomb*, 1986

A very full account of the physics, technology, and organization involved in one of the greatest of all industrial ventures, and its appalling outcome.

Rogers, E M *Physics for the Inquiring Mind*, 1960

An introduction to the general ideas of physics, with much historical detail, aimed at the non-specialist reader by an outstanding and innovating teacher; copious line drawings by the author.

Schwartz, Joseph and McGuinness, Michael *Einstein for Beginners*, 1993

A cartoon treatment of the great physicist and his ideas. This book will at least get you started, and relatively painlessly.

Segrè, E *From Falling Bodies to Radio Waves*, 1984

An historically-based account of classical physics through the achievements of such as Galileo, Newton, Faraday, Clausius, Maxwell, and Gibbs.

Snow, C P *The Physicists*, 1981

A excellent way to feel part of the 'golden age' of physics during the 1930s and to experience the moral torment of the physics community during the development and subsequent use of the atom bomb.

Squires, Euan *The Mystery of the Quantum World*, 1994

An ideal text for the general reader who wishes to develop their understanding of the philosophical issues raised by quantum theory.

Stehle, P *Order, Chaos, Order*, 1994

A largely non-technical account, with more detailed appendices, of the complex early years of the 20th century, during which the perceived weaknesses of classical physics were resolved through the invention of quantum mechanics.

Stewart, Ian *Does God Play Dice?*, 1989

An introduction to the mathematics of chaos theory without the reader needing anything other than some basic numeracy.

Sykes, Christopher (ed) *No Ordinary Genius*, 1994

An amply-illustrated tribute to Richard Feynman, Nobel prizewinner and pioneer in the understanding of quantum mechanics, who died in 1988.

Taylor, John (ed) *Tributes to Paul Dirac*, 1987

Based on papers presented at the memorial meeting for Paul Dirac, Cambridge, 19 April 1985.

Weinberg, Steven *Quarks: The stuff of matter*, 1983

An intriguing discussion of the early history of particle physics. It covers, in particular, the discoveries of the electron and the atomic nucleus by introducing a number of basic physical principles and emphasizing how they underlie our ability to 'see' within the atom.

Zohar, Danah *The Quantum Self,* 1990

An encouraging book that attempts to make the links between our understanding of atomic behaviour and our understanding of ourselves.

Zohar, Danah and Marshall, Ian *The Quantum Society*, 1994

This sequel to *The Quantum Self* carries the argument further, making links between the quantum idea and a new and better society.

6 Web Sites

Accelerator Physics Group
http://wwwslap.cern.ch/

Virtual library dedicated to accelerator physics, with pages on design and components, as well as direct links to laboratories throughout the world.

ALEPH Experiment
http://alephwww.cern.ch/

Home page of one of the four high energy particle physics experiments, which use the Large Electron Positron Collider at the European Laboratory for Particle Physics. This Web site displays images of the components used in this experiment, including a cutaway schematic diagram of the huge detector assembly. The site also allows the visitor to discover more about the physics the collider and associated instruments are used to observe.

Animated Hologrpaher
http://www.holoworld.com/holo/demo1.html

Holograms, holography, and lasers are clearly explained in this 'slide show' demonstration of how holograms are made. Each slide has an associated RealAudio commentary to help illuminate each stage of this scientific technique.

BBC Science in Action
http://www.bbc.co.uk/sia/

Well-presented site from the British Broadcasting Corporation's *Science in Action* programme. The site is divided up into five sections, 'Light', 'Air', 'Microbes', 'Mixtures', and 'Force'. Each section contains several topics. The scientific principles behind each topic are explained, followed by a series of activities and investigations. The activities require either Shockwave or Flash plug-ins to work, but they can be downloaded from the site.

Beam Me Up: Photons and Teleportation
http://www.sciam.com/explorations/122297teleport/

Part of a larger site maintained by *Scientific American*, this page reports on the amazing research conducted by physicists at the University of

Innsbruck, Austria, who have turned science fiction into reality by teleporting the properties of one photon particle to another. Learn how quantum teleportation was accomplished and find out what the likelihood is that you will soon be teleporting yourself to those distant vacation spots around the world.

Biographies of Physicists
http://hermes.astro.washington.edu/scied/physics/physbio.html

Valuable compilation of biographies of the most famous physicists of all time, including Aristotle, Da Vinci, Kepler, Galilei, Newton, Franklin, Curie, Feynman, and Oppenheimer. Be careful to bookmark it properly because you can easily get lost in the many pages of this site.

Blackett, Patrick Maynard Stuart
http://www.nobel.se/physics/laureates/1948/blackett-bio.html

Portrait of the physicist and Nobel laureate. It traces the development of his theoretical interests. There is a picture of Blackett and the text is hyperlinked to lead you to pages on related people and places.

Born, Max
http://www-history.mcs.st-and.ac.uk/history/Mathematicians/Born.html

Biography of the German-born British physicist. The Web site details the work of Born, and his relationships with his contemporaries and colleagues. Also included are several literature references for further reading on the physicist, and a portrait photograph of Born.

Colour Matters
http://www.colormatters.com/entercolormatters.html

Examination of the interaction between colour and the disciplines of physiology, psychology, philosophy, architecture, and optics.

Contributions of 20th-Century Women to Physics
http://www.physics.ucla.edu/~cwp/wipl.jpg

Not just a fanfare for female scientists, this site contains a searchable archive of physics research conducted by women in the 20th century. Other pages celebrate the work of early pioneers such as Agnes Pockels, and provide a history of early developments in areas such as nuclear physics.

Cryogenics
http://www-csa.fnal.gov/

Information site of the Cryogenic Society of America (CSA), a non-profit technical society serving all those interested in any phase of cryogenics. This site provides online access to publications, supported by details of meetings and events.

De Broglie, Louis
http://www-history.mcs.st-and.ac.uk/history/Mathematicians/
Broglie.html

Biographical details and a photograph of Louis de Broglie, the famous French physicist and mathematician. There are also links to de Broglie's most famous work on quantum mechanics and to many of his contemporaries.

Debye and Strong Electrolytes
http://dbhs.wvusd.k12.ca.us/Chem-History/
Debye-Strong-Electrolyte.html

Transcript of Peter Debye's and Erich Hüchel's paper *On The Theory Of Electrolytes: I. Freezing Point Depression and Related Phenomena*. The paper as presented on this Web site is not complete, but it does cover all the essential points of the paper.

Doppler Effect
http://www.ncsa.uiuc.edu/Cyberia/Bima/doppler.html

Explanation of the Doppler effect and links to other related sites.

Edward Teller Profile
http://www.achievement.org/autodoc/page/tel0pro-1

Description of the life and work of the 'father of the hydrogen bomb', physicist Edward Teller. The Web site contains not only a profile and biographical information, but also holds a lengthy interview with Teller from 1990 accompanied by a large number of photographs, video sequences, and audio clips.

Effect of Magnetization on the Nature of Light Emitted by a Substance
http://dbhs.wvusd.k12.ca.us/Chem-History/Zeeman-effect.html

Excerpt from Pieter Zeeman's paper published in *Nature*, vol. 55 on the 11th of February 1897. The effect strong magnetic fields have on light is described in the paper, and it is accompanied here by a photograph Zeeman himself took which illustrates the effect. Also shown is a photograph of Zeeman, who in 1926 won the Nobel Prize for physics for this work.

Einstein's Legacy

http://www.ncsa.uiuc.edu/Cyberia/NumRel/EinsteinLegacy.html

Illustrated introduction to the man and his greatest legacy – relativity and the concept of space-time. There is a video- and audio-clip version of the page courtesy of a US scientist and details about how current research is linked to Einstein's revolutionary ideas.

Einstein: Still Right After All These Years

http://whyfiles.news.wisc.edu/052einstein/

Easy-to-understand explanation of some of Einstein's theories. The explanations are grouped under several headings, such as: 'The speed of light', 'Black holes', and 'Gravity'.

Electricity and Magnetism

http://library.thinkquest.org/12632/magnetism/

Clearly presented explanation of 'electric charge', 'electro-magnetism' and other aspects relating to this area of study.

Emilio Segrè Visual Archives

http://www.aip.org/history/esva/

Part of the American Institute of Physics site, this page includes thousands of historical photographs, slides, lithographs, engravings, and other visual materials mostly featuring physicists and astronomers of the 20th century. In addition to photos, you will find stories and quotes from such well known scientists as Albert Einstein, Marie Curie, Isaac Newton, Galileo, Enrico Fermi, Neils Bohr, and many others.

EnergyEd

http://www.ergon.com.au/energyed/

Light-hearted and educational introduction to electricity and renewable energy with the opportunity to take a virtual tour around a power station.

Eric's Treasure Trove of Physics

http://www.treasure-troves.com/physics/physics.html

Searchable database of facts relating to physics compiled by Eric W Weisstein, an 'Internet Encyclopedist'. You can search by keywords or subject, and the key terms in each definition are linked to their own definition.

Essay about Marie and Pierre Curie

http://www.nobel.se/physics/articles/curie/

Extended essay on the life and deeds of Pierre and Marie Curie. The site spans the entirety of the couple's astonishing career and their contribution to the promotion of the understanding of radioactivity. It includes sections on Pierre's and Marie's joint research on radiation phenomena as well as Marie's own work after Pierre's untimely death in 1906: her discovery of the elements radium and polonium, her second Nobel Prize in Chemistry, her difficulties with the press, and her family life.

European Laboratory for Particle Physics

http://www.cern.ch/

Information about CERN, the world-class physics laboratory in Geneva. As well as presenting committees, groups, and associations hosted by the Laboratory, this official site offers important scientific material and visual evidence on the current activities and projects. Visitors will also find postings about colloquia, schools, meetings, and other services offered there. A special section is devoted to the history of the World Wide Web and the pioneering contribution of CERN in its conception and expansion.

Explanation of Temperature Related Theories

http://www.unidata.ucar.edu/staff/blynds/tmp.html

Detailed explanatory site on the laws and theories of temperature. It explains what temperature actually is, what a thermometer is, and the development of both, complete with illustrations and links to pioneers in the field. There is a temperature conversion facility and explanations of associated topics such as kinetic theory and thermal radiation.

ExploreScience.com

http://www.explorescience.com

One of many sites promising to make physics fun, ExploreScience.com contains a wide variety of interactive (using Shockwave) physics activities that might actually succeed in doing so. Modules cover GCSE-relevant topics such as electricity and magnetism, wave motion, mechanics, optics, and astronomy.

Exploring Gravity

http://www.curtin.edu.au/curtin/dept/phys-sci/gravity/index2.htm

Interactive tour of all things related to gravity. The site includes quizzes and photos and covers 'Introductory', 'Intermediate', and 'Advanced' stages of study.

Feynman, Richard Phillips

http://www-groups.dcs.st-and.ac.uk/history/Mathematicians/
Feynman.html

Devoted to the life and contributions of physicist Richard Feynman. In addition to biographical information, you will find a list of references about Feynman and links to essays that reference him. The text of this essay includes hypertext links to the essays of those mathematicians and thinkers who influenced Feynman. You will also find an image of him, which you can click on to enlarge, and a map of his birthplace.

Fibre Optic Chronology

http://www.sff.net/people/Jeff.Hecht/Chron.html

Timeline of the history of fibre optics. Starting with the discovery of glass in around 2500 bc and leading up to the late 1970s, this site details every discovery relevant to the history of the fibre optic cables which, among other things, make the Internet possible.

Free Fall and Air Resistance

http://www.glenbrook.k12.il.us/gbssci/phys/Class/newtlaws/u2l3e.html

Detailed explanation of Newton's second law of motion, accompanied by diagrams and animations to help illustrate the concepts.

From Apples to Orbits: The Gravity Story

http://library.thinkquest.org/27585/

Find out all about the force that brought that legendary fruit down on top of Isaac Newton's head, and its importance to the development of space-age science. This nicely packaged site from the impressive ThinkQuest domain considers the history of gravity theory as well as small and large-scale effects of the force, and includes sample experiments to test them out.

Fusion

http://www.pppl.gov/~rfheeter/

All-text site packed with information about fusion research and its applications. It is quite well organized and includes a glossary of commonly used terms to aid the uninitiated.

GCSE Bitesize Physics

http://www.bbc.co.uk/education/gcsebitesize/science_physics/
index.shtml

One-stop revision shop for GCSE physicists provided by the BBC as part of an all-encompassing GCSE site. Covering all the major topics of the

GCSE physics curriculum, the site offers 'revision bites' for those at the start of their pre-exam preparation, and 'test bites' for the more confident.

Guided Tour of Fermilab Exhibit

http://www.fnal.gov/pub/tour.html

Guided tour of the particle physics laboratory Fermilab in Illinois, USA. It includes an explanation of the principles of particle physics, a guide to particle accelerators, and an insight into the experiments currently being conducted at the lab.

Hawking, Stephen

http://www.norfacad.pvt.k12.va.us/project/hawking/hawking.htm

Stephen Hawking's own home page, with a brief biography, disability advice, and a selection of his lectures, including 'The beginning of time' and a series debating the nature of space and time.

Heaviside, Oliver

http://www-history.mcs.st-and.ac.uk/history/Mathematicians/ Heaviside.html

Extensive biography of the English physicist. The site contains a description of his contribution to physics, and in particular his simplification of Maxwell's 20 equations in 20 variables, replacing them by two equations in two variables. Today we call these 'Maxwell's equations' forgetting that they are in fact 'Heaviside's equations'.

Heisenberg, Werner Karl

http://www-groups.dcs.st-and.ac.uk/history/Mathematicians/ Heisenberg.html

Part of an archive containing the biographies of the world's greatest mathematicians, this site is devoted to the life and contributions of physicist Werner Heisenberg. In addition to biographical information, you will find a list of references about Heisenberg and links to other essays in the archive that reference him. The text of this essay includes hypertext links to the essays of those mathematicians and thinkers who influenced Heisenberg. You will also find an image of him, which you can click on to enlarge, and a map of his birthplace.

History of Radar

http://www.fi.edu/weather/radar/history.html

A brief history of the development of radar, which gives credit to British researchers for their role in the winning of the Battle of Britain.

How Things Work
http://howthingswork.virginia.edu/

Well-organized site providing answers to many questions relating to physics. There is a large database of previously asked questions that may be readily accessed. Run by the professor of Physics at the University of Virginia, USA, the site is an essential point of reference for anybody with an interest in physics.

How Things Work: the Physics of Everyday Life
http://rabi.phys.virgina.edu/HTW/

Questions and answers on the physics of a huge range of aspects of everyday life, from bouncing balls to roller coasters, and electronic air cleaners to nuclear weapons. These pages, set up by the University of Virginia, USA, are based on a book of the same title; although often entertaining in tone, their physics content is challenging but clearly written.

How to Make a Pinhole Camera
http://www.pinhole.com/resources/pinhole126/pinhole.htm

Full instructions, with accompanying diagrams, of how to build your own pinhole camera.

Hunting of the Quark
http://researchmag.asu.edu/stories/quark.html

Commentary on the research carried out at Arizona State University, USA, into sub-atomic particles. There is also a subsidiary page on the complexities of the subatomic world.

Interactive Physics Problem Set
http://vydra.karlov.mff.cuni.cz/HomePages/Kotrla/Berkeley/Contents.html

Nearly 100 practice problems at various levels of difficulty, with detailed answers. The subjects covered include 'Position, velocity, and acceleration', 'Forces', and 'Energy and work'.

Internet Plasma Physics Experience
http://ippex.pppl.gov/ippex/

Operate your own virtual Tokamak fusion reactor and get a hands-on feel for this exciting branch of physics. This information-packed site explains the key physics concepts involved in nuclear fusion and plasma physics, such as matter, electricity and magnetism, and energy.

Introduction to Mass Spectrometry
http://www.scimedia.com/chem-ed/ms/ms-intro.htm

Good introduction to the mass spectrometer and how it works. Different mass analyser designs are described, together with sections on ionization and ion detectors.

Introduction to Waves
http://id.mind.net/~zona/mstm/physics/waves/waves.html

Interactive site that begins with the basics – explaining and allowing you to manipulate wavelength, amplitude, and phase shift of a simple wave. Further into the site there are more complex examinations of such things as Huygen's principle, interference, and wave propagation. You will need to have a Java-enabled browser to get the most out of this site.

Isotopes and Atomic Weights
http://dbhs.wvusd.k12.ca.us/Chem-History/Aston-MassSpec.html

Transcript of part of English physicist Francis Aston's paper from 1920 outlining his ideas on atomic weights and isotopes. The paper led directly to the production of the first mass spectrometer, and thus to the separation of many isotopic elements.

Jeans, Sir James Hopwood
http://www-groups.dcs.st-and.ac.uk/history/Mathematicians/
Jeans.html

Part of an archive containing the biographies of the world's greatest mathematicians, this site is devoted to the life and contributions of James Jeans. In addition to biographical information, you will find a list of references about Jeans and links to other essays in the archive that reference him. The text of this essay includes hypertext links to the essays of those mathematicians and thinkers who influenced Jeans. You will also find an image of him, which you can click on to enlarge, and a map of his birthplace.

Kaleidoscope Heaven
http://kaleidoscopeheaven.org/

Central hub of resources for anyone interested in kaleidoscopes. This page has information on the history of the kaleidoscope, where to find materials to make your own, how to make one, a current educational project, and even an essay on the alleged health benefits of the kaleidoscope.

Landau, Lev Davidovich

http://www-groups.dcs.st-and.ac.uk/history/Mathematicians/
Landau_Lev.html

Part of an archive containing the biographies of the world's greatest mathematicians, this site is devoted to the life and contributions of theoretical physicist Lev Landau. In addition to biographical information, you will find a list of references about Landau and links to other essays in the archive that reference him. The text of this essay includes hypertext links to the essays of those mathematicians and thinkers who influenced Landau. You will also find an image of him, which you can click on to enlarge, and a map of his birthplace.

Leaping Leptoquarks

http://www.sciam.com/explorations/032497lepto/032497horgan.html

Part of a larger site maintained by *Scientific American*, this page follows the trail of two German physicists in search of the elusive leptoquark. Find out about events that led them to believe they were witnessing a new phenomenon: a particle that combined aspects of the two elementary particles that make up atoms, leptons and quarks. The text includes hypertext links to further information, and there is also a list of related links at the bottom of the page.

Learn Physics Today

http://library.thinkquest.org/10796/

Divided into three online lessons of 'Mechanics', 'Lights and waves', and 'Electricity', this is a comprehensive physics Web site. It features interactive tests to assess learning progress, and a lesson summary page giving a quick overview of the topics. An online calculator must be chosen before any of the lessons are looked at, as this is needed in some of the calculations.

Leo Szilard Home Page

http://www.dannen.com//szilard.html

Home page for the Hungarian-born physicist who was one of the first to realize the significance of splitting the atom. This site includes an illustrated biography, audio clips of people who worked with him, and documentation relating to the USA's decision to develop and then use the A-bomb.

Life and Theories of Albert Einstein

http://www.pbs.org/wgbh/nova/einstein/index.html

Heavily illustrated site on the life and theories of Einstein. There is an illustrated biographical chart, including a summary of his major achievements and their importance to science. The theory of relativity gets an understandably more in-depth coverage, along with photos and illustrations. The pages on his theories on light and time include illustrated explanations and an interactive test. There is also a 'Time-traveller' game demonstrating these theories.

Light!
http://library.thinkquest.org/28160/

Thought-provoking ThinkQuest site on the physics of light, available in English or Polish. Topics considered include lasers, fibre optics, wavelengths, and the speed of light.

Little Shop of Physics: Online Experiments
http://littleshop.physics.colostate.edu/Experiments.html

Unusual, easy-to-follow site divided into experiments you can do in three ways – 'Using common household items', 'You can do using your computer', and 'Using the Shockwave plug-in'.

Look Inside the Atom
http://www.aip.org/history/electron/jjhome.htm

Part of the American Institute of Physics site, this page examines J J Thomson's 1897 experiments that led to the discovery of a fundamental building block of matter, the electron. The site includes images and quotes and a section on the legacy of his discovery. Also included is a section on suggested readings and related links.

Los Alamos
http://www.lanl.gov/external/welcome/

Web site of the institution that gave the world nuclear weapons. A small history section recalls the early days of the laboratory but the information here is mostly about the current work of this important scientific institution.

Making Waves
http://www.smgaels.org/physics/home.htm

Easy-to-follow site created by students at a US high school. The topics covered include 'Sound', 'Microwave', 'Light and lasers', and 'Gamma rays'.

Maths and Physics Help Page
http://www2.ncsu.edu/unity/lockers/users/f/felder/public/kenny/home.html

Demystifying papers such as 'Think like a physicist' and 'The day the universe went all funny: an introduction to special relativity'.

Mechanical Properties of Plastics
http://www.stemnet.nf.ca/DeptEd/intermediate/production/
02bigideas/01properties/c01properties/plasticmech.htm

Brief, but clear description of the mechanical properties of plastics, including the properties of a plastic when subjected to a physical force. The page explains concepts such as creep and plastic memory.

NetScience: Physics
http://library.thinkquest.org/3616/physics/

Clearly-presented site aimed at high school students. The site covers vectors, motion, forces, momentum, work, energy, and power.

Niels Bohr Institute History
http://www.nbi.dk/nbi-history.html

Short survey of the development of the Niels Bohr Institute, so intimately connected with Bohr's life work as a physicist. There are sections on the Institute today, a history from 1929–65, and a picture gallery.

Nuclear Energy: Frequently Asked Questions
http://www-formal.stanford.edu/jmc/progress/nuclear-faq.html

Answers to the most commonly asked questions about nuclear energy, particularly with a view to sustaining human progress. It contains many links to related pages and is a personal opinion that openly asks for comment from visitors.

Nuclear Physics
http://www.scri.fsu.edu/~jac/Nuclear/

'Hyper-textbook' of nuclear physics, with an introduction that includes a graphical description of the size and shape of nuclei and their other properties. The site also includes information about the work of nuclear physicists, and the uses and applications of nuclear physics from medicine, through energy, to smoke detectors.

Particle Adventure
http://particleadventure.org/frameless/index.html

Comprehensive site about particles. It covers many topics including what the world is made of, what holds it together, and decay and annihilations. The subject matter is difficult in places, but the introductions to each topic may be of use to GCSE pupils, and the site is of interest to all those who wish to stretch themselves.

Particle Physics

http://www.pparc.ac.uk/

Particle Physics and Astronomy Research Council (PPARC) Web site that includes information about its extensive research programmes in the fields of astronomy, planetary science, and particle physics, as well as more general information on stars and the origins of the universe.

Pauli, Wolfgang

http://www-groups.dcs.st-and.ac.uk/history/Mathematicians/
Pauli.html

Part of an archive containing the biographies of the world's greatest mathematicians, this site is devoted to the life and contributions of physicist Wolfgang Pauli. In addition to biographical information, you will find a list of references about Pauli and links to other essays in the archive that reference him. The text of this essay includes hypertext links to the essays of those mathematicians and thinkers who influenced Pauli. You will also find an image of him, which you can click on to enlarge, and a map of his birthplace.

Physics 2000

http://www.colorado.edu/physics/2000/

Stylish site on physics covering a range of topics under the four main headings of 'What is it?', 'Einstein's legacy', 'The atomic lab', and 'Science trek'. Each one is well illustrated and makes good use of interactive Java applets, bringing complex ideas to life.

Physics-related Information

http://www.alcyone.com/max/physics/index.html

Excellent revision resource for GCSE physicists and above. This Web site contains explanations of and proofs for some of the major laws of physics, a guide to orders of magnitude, and links to some good related sites.

PhysicsTutor.com

http://www.physicstutor.com/newcat.html

Search for help with physics problems with this online 'teacher'. Select a topic and browse the previous questions, or type in a key word to bring up a related question.

Physics Zone

http://www.sciencejoywagon.com/physicszone/

Searchable 'resource for learning introductory level, algebra-based physics' arranged by topic. The site uses text, animation, slide shows, and interactive (Shockwave) 'labs' to explore 'rotational and circular stuff', electricity, motion, and more.

Planck, Max Karl Ernst Ludwig

http://www-history.mcs.st-and.ac.uk/history/Mathematicians/
Planck.html

Biography of the eminent German physicist Max Planck. Planck is thought of by many to have been more influential than any other in the foundation of quantum physics, and partly as a consequence has a fundamental constant named after him. This Web site contains a description of his working relationships with his contemporaries, and also features a photograph of Planck. Several literature references for further reading on the works of Planck are also listed.

Quantum Age Begins

http://www-history.mcs.st-and.ac.uk/history/HistTopics/
The_Quantum_age_begins.html

St Andrews University-run Web site chronicling the discovery of quantum theory. Biographical details of the mathematicians and physicists involved are also provided. The site also includes links to many other history of mathematics-related Web resources.

Radiation and Radioactivity

http://www.rso.utah.edu/train/cover.htm

Lesson on radioactivity. It begins with a definition of radioactivity, and then covers the atom and types of radiation including alpha, beta, and gamma radiation and X-rays, plus the properties of each. A well presented lesson with clear, helpful diagrams.

Radiation Reassessed

http://whyfiles.news.wisc.edu/020radiation/index.html

Part of the Why Files project published by the National Institute for Science Education (NISE) and funded by the National Science Foundation, this page provides insight into the controversy concerning the health effects of ionizing radiation. The highly readable text, laid out over 12 pages, includes information about what radiation studies from Hiroshima, Nagasaki, and Chernobyl have taught us, and what scientists are saying about exposure to small amounts of radiation. Is it always harmful? Numerous images and diagrams enhance the text throughout. The more tricky terms are linked to a glossary, and you will find a comprehensive bibliography of sources for further research.

Röntgen, Wilhelm Conrad

http://www.nobel.se/physics/laureates/1901/rontgen-bio.html

Biography of Wilhelm Conrad Röntgen, the German physicist who first realized the huge potential of the electromagnetic field of X-rays. The presentation includes sections on Röntgen's early years and education, his academic career and scientific experiments, and the miraculous coincidences that led him to his great discovery of X-rays.

Rutherford on the Discovery of Alpha and Beta Radiation

http://dbhs.wvusd.k12.ca.us/Chem-History/
Rutherford-Alpha&Beta.html

Transcript of Rutherford's paper 'Uranium radiation and the electrical conduction produced by it' which first appeared in the *Philosophical Magazine* in January 1899. The full paper with diagrams is reproduced here, and in it Rutherford describes the nature of the two types of radiation he discovered to be emitted from uranium as it decays.

Rutherford's Discovery of Half-Life

http://dbhs.wvusd.k12.ca.us/Chem-History/Rutherford-half-life.html

Transcript of Ernest Rutherford's paper describing his discovery of the half life of radioactive materials. The paper entitled 'A radioactive substance emitted from thorium compounds' first appeared in the *Philosophical Magazine* in January 1900. The reproduction of it on this Web site is complete with several diagrams and all the relevant equations.

Sakharov, Andrei

http://www.nobel.se/peace/laureates/1975/sakharov-autobio.html

Brief autobiography written by the physicist and human rights campaigner for the Nobel Committee after he won the 1975 Peace Prize. Sakharov describes how his upbringing inculcated a love of literature, science, and music. He relates his pioneering work in Soviet nuclear physics and the deepening moral doubts this aroused. He movingly describes how it was his conscience that propelled him into an unwilling engagement with politics.

Schrödinger, Erwin

http://www-groups.dcs.st-and.ac.uk/history/Mathematicians/
Schrodinger.html

Part of an archive containing the biographies of the world's greatest mathematicians, this site is devoted to the life and contributions of physicist Erwin Schrödinger. In addition to biographical information, you will find a list of references about Schrödinger and links to other essays in the archive that reference him. The text of this essay includes hypertext links to the essays of those mathematicians and thinkers who influenced Schrödinger. You will also find an image of him, which you can click on to enlarge, and a map of his birthplace.

Schrödinger's Cation
http://www.sciam.com/explorations/061796explorations.html

Part of a larger site maintained by *Scientific American*, this page features an explanation of the quantum mechanics paradox known as 'Schrödinger's cat', a thought experiment devised by Erwin Schrödinger to illustrate the difference between the quantum and macroscopic worlds. You will also find information about another experiment that showed an atom actually existing in two states at one time. The text includes hypertext links to further information about quantum mechanics, and there is also a list of related links at the bottom of the page.

Static Electricity
http://www.sciencemadesimple.com/static.html

Excellent site, explaining the theory simply, but thoroughly. It also includes several easy-to-do experiments to demonstrate the theory behind static electricity.

Stephen Hawking's Universe
http://www.wnet.org/archive/hawking/html/home.html

Interesting site with original essays by some of today's leading figures in cosmology. The topics covered include 'Strange stuff explained' and 'Unsolved mysteries'.

Subatomic Logic
http://www.sciam.com/explorations/091696explorations.html

Part of a larger site maintained by *Scientific American*, this page provides information on the recent progress of scientists who are attempting to harness quantum physics to run a lightning fast, super-charged 'quantum computer'. This article explains how a quantum computer would work and why it would be so much faster than silicon-based computer systems. The text includes hypertext links to further information, and there is also a list of related links.

Tesla, Nikola
http://www.neuronet.pitt.edu/~bogdan/tesla/

Short biography of the electrical inventor plus quotations by and about Tesla, anecdotes, a photo gallery, and links to other sites of interest.

Unit Converter
http://www.webcom.com/legacysy/convert2/convert2.html

Simple, but very effective site that allows conversion from values in any of an astonishingly wide variety of units to any other relevant unit. The units available are listed together in groups from 'Acceleration', through 'Luminance' and 'Specific heat', to 'Weight (mass)'. All you have to do is select the category, the units you want to convert from and into, and type in the value you want to convert to get a fairly immediate answer. Please note: you will need a Java-enabled browser for this site to work.

Usenet Relativity FAQ
http://math.ucr.edu/home/baez/physics/relativity.html

Concise answers to some of the most common questions about relativity. The speed of light and its relation to mass, dark matter, black holes, time travel, and the Big Bang are some of the things covered by this illuminating series of articles based both on Usenet discussions and good reference sources. The site also directs the visitors to appropriate discussion groups where they can pose more questions, and also solicits more articles on themes not yet covered by the 'Frequently Asked Questions'.

Visual Physics
http://www.visualphysics.com/

Downloadable programs for teaching and learning about physics, using visual and interactive techniques. The topics covered include functions, energy, momentum, waves, and aspects of mechanics.

Wester's Online Guide to Physics
http://www.msms.doe.k12.ms.us/ap_physics/

Clearly-presented tutorial including lessons, example problems to test your knowledge, and a 'Frequently Asked Questions'.

7   Glossary

A
symbol for ampere, a unit of electrical current.

aberration, optical
any of a number of defects that impairs the image in an optical instrument. Aberration occurs because of minute variations in lenses and mirrors, and because different parts of the light spectrum are reflected or refracted by varying amounts.

absolute zero
lowest temperature theoretically possible according to kinetic theory, zero kelvin (0 K), equivalent to $-273.15°C/-459.67°F$, at which molecules are in their lowest energy state. Although the third law of thermodynamics indicates the impossibility of reaching absolute zero, in practice temperatures of less than a billionth of a degree above absolute zero have been produced. Near absolute zero, the physical properties of some materials change substantially; for example, some metals lose their electrical resistance and become superconducting.

absorption
taking up of matter or energy of one substance by another, such as a liquid by a solid (ink by blotting paper) or a gas by a liquid (ammonia by water). In physics, absorption is the phenomenon by which a substance retains the energy of radiation of particular wavelengths; for example, a piece of blue glass absorbs all visible light except the wavelengths in the blue part of the spectrum; it also refers to the partial loss of energy resulting from light and other electromagnetic waves passing through a medium. In nuclear physics, absorption is the capture by elements, such as boron, of neutrons produced by fission in a reactor.

absorption of light and heat
The energy of electromagnetic waves, for example light and heat radiation, is absorbed in discrete amounts, called quanta, and enough quanta may be absorbed by certain materials to allow photochemical reactions to take place. As well as absorption, some radiation may be transmitted with more or less disturbance of direction and some may be reflected. A particular substance may absorb rays of certain frequencies or wavelengths, allowing the others to be transmitted or reflected. Long waves, such as radio waves, can pass through optically opaque obstacles with very little absorption. Shorter heat waves are absorbed readily by dark substances such as lamp black, whilst the various wavelengths that correspond to the different colour sensations are variously affected by different substances, the result determining the colour as seen by

transmitted light. Absorption and emission of electromagnetic radiation, is the basis of spectroscopic methods employed to give information on the geometry and electronic structures of molecules. Molecules have distinct rotational, vibrational, and electronic energy levels, and can absorb electromagnetic radiation from a source. The absorption is quantized, that is, only an exact amount of energy will raise the molecule to its next energy level. It is possible to produce an absorption spectrum showing which wavelengths have been absorbed, and since different molecules and parts of molecules have different wavelength requirements, this serves as an analytical tool.

AC

abbreviation for *alternating current.*

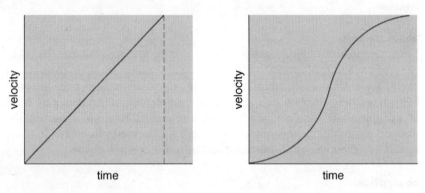

Acceleration can be depicted graphically by mapping velocity against time. Where acceleration is constant the graph is a straight line.

acceleration

rate of change of the velocity of a moving body. It is usually measured in metres per second per second (m s^{-2}) or feet per second per second (ft s^{-2}). Because velocity is a vector quantity (possessing both magnitude and direction) a body travelling at constant speed may be said to be accelerating if its direction of motion changes. According to Newton's second law of motion, a body will accelerate only if it is acted upon by an unbalanced, or resultant, force. Acceleration due to gravity is the acceleration of a body falling freely under the influence of the Earth's gravitational field; it varies slightly at different latitudes and altitudes. The value adopted internationally for gravitational acceleration is 9.806 m s^{-2}/32.174 ft s^{-2}.

The average acceleration a of an object travelling in a straight line over a period of time t may be calculated using the formula:

$$a = \frac{\text{change of velocity}}{t}$$

or, where u is its initial velocity and v its final velocity:

$$a = \frac{(v - u)}{t}$$

A negative answer shows that the object is slowing down (decelerating). See also equations of motion.

accelerator

device to bring charged particles (such as protons and electrons) up to high speeds and energies, at which they can be of use in industry, medicine, and pure physics. At low energies, accelerated particles can be used to produce the image on a television screen and generate X-rays (by means of a cathode-ray tube), destroy tumour cells, or kill bacteria. When high-energy particles collide with other particles, the fragments formed reveal the nature of the fundamental forces.

The first accelerators used high voltages (produced by Van de Graaff generators) to generate a strong, unvarying electric field. Charged particles were accelerated as they passed through the electric field. However, because the voltage produced by a generator is limited, these accelerators were replaced by machines where the particles passed through regions of alternating electric fields, receiving a succession of small pushes to accelerate them. The first of these accelerators was the *linear accelerator* or *linac*. The linac consists of a line of metal tubes, called drift tubes, through which the particles travel. The particles are accelerated by electric fields in the gaps between the drift tubes. Another way of making repeated use of an electric field is to bend the path of a particle into a circle so that it passes repeatedly through the same electric field. The first accelerator to use this idea was the *cyclotron* pioneered in the early 1930s by US physicist Ernest Lawrence. One of the world's most powerful accelerators is the 2 km/1.25 mi diameter machine at Fermilab, Illinois, USA. This machine, the Tevatron, accelerates protons and antiprotons and then collides them at energies up to a thousand billion electron volts (or 1 TeV, hence the name of the machine). The largest accelerator is the Large Electron Positron Collider at CERN, Switzerland, operational 1989–2000, which has a circumference of 27 km/16.8 mi around which electrons and positrons are accelerated before being allowed to collide. The world's longest linac is also a colliding beam machine: the Stanford Linear Collider, in California, USA, in which electrons and positrons are accelerated along a straight track, 3.2 km/2 mi long, and then steered to a head-on collision with other particles, such as protons and neutrons. Such experiments have been instrumental in revealing that protons and neutrons are made up of smaller elementary particles called quarks.

A cyclotron consists of an electromagnet with two hollow metal semicircular structures, called dees, supported between the poles of an electromagnet. Particles such as protons are introduced at the centre of the machine and travel outwards in a spiral path, being accelerated by an oscillating electric field each

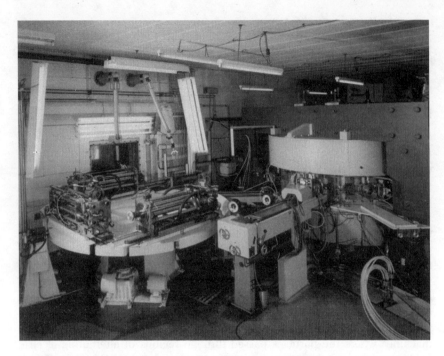

A cyclotron accelerator that employs a rotating, water-cooled target to produce isotopes. These include gallium-67 (half-life 78 hours) and indium-111 (half-life 67 hours). Sodium-22, cobalt-57, and cadmium-109 (all with half-lives in excess of 150 hours) are made in irradiations extending for up to 150 hours. AEA Technology

time they pass through the gap between the dees. Cyclotrons can accelerate particles up to energies of 25 MeV (25 million electron volts); to produce higher energies, new techniques are needed. In the synchrotron, particles travel in a circular path of constant radius, guided by electromagnets. The strengths of the electromagnets are varied to keep the particles on an accurate path. Electric fields at points around the path accelerate the particles. Early accelerators directed the particle beam onto a stationary target; large modern accelerators usually collide beams of particles that are travelling in opposite directions. This arrangement doubles the effective energy of the collision.

accumulator

storage battery – that is, a group of rechargeable secondary cells. A familiar example is the lead–acid car battery.

An ordinary 12-volt car battery consists of six lead–acid cells which are continually recharged when the motor is running by the car's alternator or dynamo. It has electrodes of lead and lead oxide in an electrolyte of sulphuric acid. Another common type of accumulator is the 'nife' or NiFe cell, which has electrodes of nickel and iron in a potassium hydroxide electrolyte.

The current that can be obtained from a cell depends on the active area of the plates in contact with the electrolyte. If more current is required than can be obtained from one cell, several cells are connected in parallel.

acoustics

experimental and theoretical science of sound and its transmission; in particular, that branch of the science that has to do with the phenomena of sound in a particular space such as a room or theatre. In architecture, the sound-reflecting character of an internal space.

Acoustic engineering is concerned with the technical control of sound, and involves architecture and construction, studying control of vibration, soundproofing, and the elimination of noise. It also includes all forms of sound recording and reinforcement, the hearing and perception of sounds, and hearing aids.

Sound energy spreads as vibrations in the form of pressure waves that are absorbed by soft objects such as drapery and human bodies, and reflected by hard surfaces such as walls and ceilings. These reflections are known as echoes. In a well-designed auditorium the echoes bouncing around all hard surfaces arrive so frequently at the ear, finally dying down, that the listener registers them merely as a slight extension to the original sound.

adiabatic

process that occurs without loss or gain of heat, especially the expansion or contraction of a gas in which a change takes place in the pressure or volume, although no heat is allowed to enter or leave. Adiabatic processes can be both nonreversible and approximately reversible.

adsorption

taking up of a gas or liquid at the surface of another substance, most commonly a solid (for example, activated charcoal adsorbs gases). It involves molecular attraction at the surface, and should be distinguished from absorption (in which a uniform solution results from a gas or liquid being incorporated into the bulk structure of a liquid or solid).

There are two types of adsorption of gases by solids: physical adsorption and chemisorption. *Physical adsorption* takes place if essentially physical forces, such as condensation, hold the gas to the solid. In these cases the heat of adsorption is less than 40 kJ mol^{-1} and adsorption is only appreciable at temperatures below the boiling point of the adsorbate. Physical adsorption is more a function of the adsorbate than the adsorbent, and no activation energy is involved in the process. The converse is true for *chemisorption*. As a rule chemical bonds play a part in the process. The heat of adsorption is greater than 80 kJ mol^{-1}, the process occurs at high temperatures, the adsorbent itself is significant, and activation energy may be involved.

aerodynamics

branch of fluid physics that studies the forces exerted by air or other gases in motion. Examples include the airflow around bodies moving at speed through the atmosphere (such as land vehicles, bullets, rockets, and aircraft), the behaviour of gas in engines and furnaces, air conditioning of buildings, the deposition of snow, the operation of air-cushion vehicles (hovercraft), wind loads on buildings and bridges, bird and insect flight, musical wind instruments, and meteorology. For maximum efficiency, the aim is usually to design the shape of an object to produce a streamlined flow, with a minimum of turbulence in the moving air. The behaviour of aerosols or the pollution of the atmosphere by foreign particles are other aspects of aerodynamics.

air

Air has a density at sea level of 1.283 kg/m^3. This falls off with increasing altitude to about one-tenth as much at 64,000 m, one-hundredth at 146,000 m, and only one-millionth at 293,000 m. Air is also viscous, so that a solid body moving through it experiences drag not only from the direct displacement of the air molecules but also from the sheer resistance as the molecules slide over each other.

subsonic speeds

At relatively low speeds aerodynamic drag forces are proportional to the local air density, the dimensions of the surface on which the air acts, and the square of the velocity difference between the air and the surface; thus, doubling the size of the surface doubles the forces, whereas doubling the speed multiplies the forces by four. According to Bernoulli's principle, the total energy in a given fluid flow remains everywhere constant; thus if the velocity is increased the pressure is correspondingly decreased. For this reason air flowing at less than the speed of sound through a pipe of varying diameter will have maximum velocity but minimum pressure at the points of minimum diameter and vice versa. For the same reason the acceleration of the air over the rounded leading edge of an aeroplane wing and across the curved upper surface results in a substantial reduction in pressure which, combined with a slight increase in pressure on the undersurface, lifts the aircraft.

supersonic speeds

At supersonic speeds air passing through a pipe of varying cross section behaves in the reverse manner: a reduction in duct diameter slows the airflow down and increases its pressure, while the thrust chamber of a rocket engine terminates in a diverging bell mouth to accelerate the supersonic gas and reduce its pressure. A body moving with supersonic speed cannot send signals through the atmosphere ahead of it; the disturbances that it creates can move away only sideways or to the rear. The forward limit of disturbances is a very precisely defined boundary, only about 2×10^{-8} m thick, called a shock wave. Upstream of a shock wave the airflow is always supersonic. As the air passes through the shock wave it experiences an essentially instantaneous rise in pressure, density, and temperature. Energy is transferred to the air via the wave, and this results in a considerable increase in the backwards drag force experienced as a solid body accelerates through the speed of sound. The wave form of a shock wave is unlike an ordinary sound wave in that it is not sinusoidal but flat fronted so that it instantly reaches its maximum amplitude. Passage of a shock wave is heard as a sharp crack (small amplitude, as from a whip), a sudden bang (from a pistol), or one or more heavy booms (large amplitude, from thunder or an aeroplane). At supersonic speeds the density of the air has to be taken into account, and flows may undergo violent changes in compressibility and temperature. Above Mach 5 (five times the speed of sound) these effects become so extreme that the gas molecules themselves may dissociate into atoms in an ionized state. The laws of a 'perfect gas' no longer apply at such speeds and such flow is termed hypersonic.

hypersonic speeds
Whereas the most efficient of supersonic aircraft shapes are slender, thin-winged, and provided with sharp edges and pointed noses, the most efficient hypersonic shapes are blunt wedges with rounded noses and in some cases no wings at all. Excessive kinetic heating constrains hypersonic vehicles to fly relatively high; in fact, a safe 'corridor' exists for such vehicles, with a lower boundary determined by structural temperature and an upper boundary set by the available lift.

afterimage
persistence of an image on the retina of the eye after the object producing it has been removed. This leads to persistence of vision, a necessary phenomenon for the illusion of continuous movement in films and television. The term is also used for the persistence of sensations other than vision.

alpha particle
positively charged, high-energy particle emitted from the nucleus of a radioactive atom. It is one of the products of the spontaneous disintegration of radioactive elements (see radioactivity) such as radium and thorium, and is identical with the nucleus of a helium atom – that is, it consists of two protons and two neutrons. The process of emission, *alpha decay*, transforms one element into another, decreasing the atomic (or proton) number by two and the atomic mass (or nucleon number) by four.

Because of their large mass alpha particles have a short range of only a few centimetres in air, and can be stopped by a sheet of paper. They have a strongly ionizing effect (see ionizing radiation) on the molecules that they strike, and are therefore capable of damaging living cells. Alpha particles travelling in a vacuum are deflected slightly by magnetic and electric fields.

alternating current (AC)
electric current that flows for an interval of time in one direction and then in the opposite direction, that is, a current that flows in alternately reversed

directions through or around a circuit. Electric energy is usually generated as alternating current in a power station, and alternating currents may be used for both power and lighting.

The advantage of alternating current over direct current (DC), as from a battery, is that its voltage can be raised or lowered economically by a transformer: high voltage for generation and transmission, and low voltage for safe utilization. Railways, factories, and domestic appliances, for example, use alternating current.

alternator
electricity generator that produces an alternating current.

ammeter
instrument that measures electric current (flow of charge per unit time), usually in amperes, through a conductor. It should not to be confused with a voltmeter, which measures potential difference between two points in a circuit. The ammeter is placed in series (see series circuit) with the component through which current is to be measured, and is constructed with a low internal resistance in order to prevent the reduction of that current as it flows through the instrument itself. Hot-wire, moving-iron, and dynamometer ammeters can be used for both DC and AC.

ampere
SI unit (symbol A) of electrical current. Electrical current is measured in a similar way to water current, in terms of an amount per unit time; one ampere (amp) represents a flow of one coulomb per second, which is about 6.28×10^{18} electrons per second.

The ampere is defined as the current that produces a specific magnetic force between two long, straight, parallel conductors placed 1m/3.3 ft apart in a vacuum. It is named after the French scientist André Ampère.

Ampère's rule
rule developed by French physicist André-Marie Ampère connecting the direction of an electric current and its associated magnetic currents. It states that around a wire carrying a current towards the observer, the magnetic field curls in the anticlockwise direction. This assumes the conventional direction of current flow (from the positive to the negative terminal).

amplifier
electronic device that increases the strength of a signal, such as a radio signal. The ratio of output signal strength to input signal strength is called the gain of the amplifier. As well as achieving high gain, an amplifier should be free from distortion and able to operate over a range of frequencies. Practical amplifiers are usually complex circuits, although simple amplifiers can be built from single transistors or valves.

amplitude modulation (AM)

method by which radio waves are altered for the transmission of broadcasting signals. AM waves are constant in frequency, but the amplitude of the transmitting wave varies in accordance with the signal being broadcast.

analogue signal

current or voltage that conveys or stores information, and varies continuously in the same way as the information it represents. Analogue signals are prone to interference and distortion.

The bumps in the grooves of a vinyl record form a mechanical analogue of the sound information stored, which is then converted into an electrical analogue signal by the record player's pick-up device.

analogue-to-digital converter (ADC)

electronic circuit that converts an analogue signal into a digital one. Such a circuit is needed to convert the signal from an analogue device into a digital signal for input into a computer. For example, many sensors designed to measure physical quantities, such as temperature and pressure, produce an analogue signal in the form of voltage and this must be passed through an ADC before computer input and processing. A digital-to-analogue converter performs the opposite process.

anemometer

device for measuring wind speed and liquid flow. The most basic form, the *cup-type anemometer*, consists of cups at the ends of arms, which rotate when the wind blows. The speed of rotation indicates the wind speed.

Vane-type anemometers have vanes, like a small windmill or propeller, that rotate when the wind blows. *Pressure-tube anemometers* use the pressure generated by the wind to indicate speed. The wind blowing into or across a tube develops a pressure, proportional to the wind speed, that is measured by a manometer or pressure gauge. *Hot-wire anemometers* work on the principle that the rate at which heat is transferred from a hot wire to the surrounding air is a measure of the air speed. Wind speed is determined by measuring either the electric current required to maintain a hot wire at a constant temperature, or the variation of resistance while a constant current is maintained.

anion

ion carrying a negative charge. During electrolysis, anions in the electrolyte move towards the anode (positive electrode).

An electrolyte, such as the salt zinc chloride ($ZnCl_2$), is dissociated in aqueous solution or in the molten state into doubly charged Zn^{2+} zinc cations and Cl^- anions. During electrolysis, the zinc cations flow to the cathode (to become discharged and liberate zinc metal) and the chloride anions flow to the anode (to become discharged and form chlorine gas).

annihilation
in nuclear physics, a process in which a particle and its 'mirror image' particle called an antiparticle collide and disappear, with the creation of a burst of energy. The energy created is equivalent to the mass of the colliding particles in accordance with the mass–energy equation. For example, an electron and a positron annihilate to produce a burst of high-energy X-rays. Not all particle–antiparticle interactions result in annihilation; the exception concerns the group called mesons, which belong to the class of particles that are composed of quarks and their antiquarks. See antimatter.

anode
positive electrode of an electrolytic cell, towards which negative particles (anions), usually in solution, are attracted.

antimatter
form of matter in which most of the attributes (such as electrical charge, magnetic moment, and spin) of elementary particles are reversed. Such particles (antiparticles) can be created in particle accelerators, such as those at CERN in Geneva, Switzerland, and at Fermilab in the USA. In 1996 physicists at CERN created the first atoms of antimatter: nine atoms of antihydrogen survived for 40 nanoseconds.

antiparticle
particle corresponding in mass and properties to a given elementary particle but with the opposite electrical charge, magnetic properties, or coupling to other fundamental forces. For example, an electron carries a negative charge whereas its antiparticle, the positron, carries a positive one. When a particle and its antiparticle collide, they destroy each other, in the process called 'annihilation', their total energy being converted to lighter particles and/or photons. A substance consisting entirely of antiparticles is known as antimatter.

Other antiparticles include the negatively charged antiproton and the antineutron.

apparent depth
depth that a transparent material such as water or glass appears to have when viewed from above. This is less than its real depth because of the refraction that takes place when light passes into a less dense medium. The ratio of the real depth to the apparent depth of a transparent material is equal to its refractive index.

Appleton layer or F layer
band containing ionized gases in the Earth's upper atmosphere, at a height of 150–1,000 km/94–625 mi, above the E layer (formerly the Kennelly–Heaviside layer). It acts as a dependable reflector of radio signals as it is not affected

by atmospheric conditions, although its ionic composition varies with the sunspot cycle.

The Appleton layer has the highest concentration of free electrons and ions of the atmospheric layers. It is named after the English physicist Edward Appleton.

atmosphere or standard atmosphere
in physics, a unit (symbol atm) of pressure equal to 760 mmHg, 1013.25 millibars, or 1.01325×10^5 pascals, or newtons per square metre. The actual pressure exerted by the atmosphere fluctuates around this value, which is assumed to be standard at sea level and 0°C/32°F, and is used when dealing with very high pressures.

atom (Greek *atomos* 'undivided')
smallest unit of matter that can take part in a chemical reaction, and which cannot be broken down chemically into anything simpler. An atom is made up of protons and neutrons in a central nucleus surrounded by electrons (see atomic structure). The atoms of the various elements differ in atomic number, relative atomic mass, and chemical behaviour.

Atoms are much too small to be seen by even the most powerful optical microscope (the largest, caesium, has a diameter of 0.0000005 mm/0.00000002 in), and they are in constant motion. However, modern electron microscopes, such as the scanning tunnelling microscope (STM) and the atomic force microscope (AFM), can produce images of individual atoms and molecules.

atomic energy
another name for *nuclear energy*.

atomic mass
see relative atomic mass.

atomic mass unit or dalton
unit of mass (symbol u) that is used to measure the relative mass of atoms and molecules. It is equal to one-twelfth of the mass of a carbon-12 atom, which is approximately the mass of a proton or 1.66×10^{-27} kg. The relative atomic mass of an atom has no units; thus oxygen-16 has an atomic mass of 16 daltons but a relative atomic mass of 16.

atomic number or proton number
number (symbol Z) of protons in the nucleus of an atom. It is equal to the positive charge on the nucleus. In a neutral atom, it is also equal to the number of electrons surrounding the nucleus. The chemical elements are arranged in the periodic table of the elements according to their atomic number. See also nuclear notation.

atomic radiation
energy given out by disintegrating atoms during radioactive decay, whether natural or synthesized. The energy may be in the form of fast-moving particles,

known as alpha particles and beta particles, or in the form of high-energy electromagnetic waves known as gamma radiation. Overlong exposure to atomic
radiation can lead to radiation sickness.

Radiation biology studies the effect of radiation on living organisms.
Exposure to atomic radiation is linked to chromosomal damage, cancer, and,
in laboratory animals at least, hereditary disease.

atomic structure
internal structure of an atom.

the nucleus
The core of the atom is the *nucleus*, a dense body only one ten-thousandth the
diameter of the atom itself. The simplest nucleus, that of hydrogen, comprises
a single stable positively charged particle, the *proton*. Nuclei of other elements
contain more protons and additional particles, called *neutrons*, of about the
same mass as the proton but with no electrical charge. Each element has its
own characteristic nucleus with a unique number of protons, the atomic
number. The number of neutrons may vary. Where atoms of a single element
have different numbers of neutrons, they are called isotopes. Although some
isotopes tend to be unstable and exhibit radioactivity, they all have identical
chemical properties.

electrons
The nucleus is surrounded by a number of moving *electrons*, each of which has
a negative charge equal to the positive charge on a proton, but which has a
mass of only $1/1,836$ times as much. In a neutral atom, the nucleus is
surrounded by the same number of electrons as it contains protons. According
to quantum theory, the position of an electron is uncertain; it may be found at
any point. However, it is more likely to be found in some places than others.
The region of space in which an electron is most likely to be found is called an
orbital (see orbital, atomic). The chemical properties of an element are determined by the ease with which its atoms can gain or lose electrons.

attraction and repulsion
Atoms are held together by the electrical forces of attraction between each negative electron and the positive protons within the nucleus. The latter repel one
another with enormous forces; a nucleus holds together only because an even
stronger force, called the *strong nuclear force*, attracts the protons and neutrons
to one another. The strong force acts over a very short range – the protons and
neutrons must be in virtual contact with one another (see forces, fundamental).
If, therefore, a fragment of a complex nucleus, containing some protons,
becomes only slightly loosened from the main group of neutrons and protons,
the natural repulsion between the protons will cause this fragment to fly apart
from the rest of the nucleus at high speed. It is by such fragmentation of atomic
nuclei (nuclear fission) that nuclear energy is released.

subatomic particles
High-energy physics research has discovered the existence of subatomic particles (see particle physics) other than the proton, neutron, and electron. More

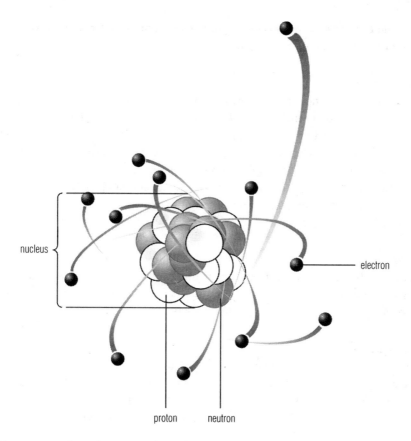

The structure of a sodium atom. The nucleus is composed of 12 protons and 11 neutrons. Eleven electrons orbit the nucleus in 3 orbits: 2 in the inner orbit, 8 in the middle, and 1 in the outer.

than 300 kinds of particle are now known, and these are classified into several classes according to their mass, electric charge, spin, magnetic moment, and interaction. The elementary particles, which include the electron, are indivisible and may be regarded as the fundamental units of matter; the *hadrons*, such as the proton and neutron, are composite particles made up of either two or three elementary particles called quarks.

background radiation
radiation that is always present in the environment. By far the greater proportion (87%) of it is emitted from natural sources. Alpha and beta particles, and gamma radiation are radiated by the traces of radioactive minerals that occur naturally in the environment and even in the human body, and by radioactive gases such as radon, which are found in soil and may seep upwards into buildings. Radiation from space (cosmic radiation) also contributes to the background level.

The *background count* is the count registered on a Gieger counter when no other radioactive source is nearby.

barometer

instrument that measures atmospheric pressure as an indication of weather. Most often used are the *mercury barometer* and the *aneroid barometer*.

In a mercury barometer a column of mercury in a glass tube, roughly 0.75 m/2.5 ft high (closed at one end, curved upwards at the other), is balanced by the pressure of the atmosphere on the open end; any change in the height of the column reflects a change in pressure. In an aneroid barometer, a shallow cylindrical metal box containing a partial vacuum expands or contracts in response to changes in pressure.

baryon

heavy subatomic particle made up of three indivisible elementary particles called quarks. The baryons form a subclass of the hadrons and comprise the nucleons (protons and neutrons) and hyperons.

becquerel

SI unit (symbol Bq) of radioactivity, equal to one radioactive disintegration (change in the nucleus of an atom when a particle or ray is given off) per second.

The becquerel is much smaller than the previous standard unit, the curie $(3.7 \times 10^{10}$ Bq). It is named after French physicist Henri Becquerel.

beta decay

disintegration of the nucleus of an atom to produce a beta particle, or high-speed electron, and an electron antineutrino. During beta decay, a neutron in the nucleus changes into a proton, thereby increasing the atomic number by one while the mass number stays the same. The mass lost in the change is converted into kinetic (movement) energy of the beta particle. Beta decay is caused by the weak nuclear force, one of the fundamental forces of nature operating inside the nucleus.

beta particle

electron ejected with great velocity from a radioactive atom that is undergoing spontaneous disintegration. Beta particles do not exist in the nucleus but are created on disintegration, beta decay, when a neutron converts to a proton by emitting an electron.

Beta particles are more penetrating than alpha particles, but less so than gamma radiation; they can travel several metres in air, but are stopped by 2–3 mm of aluminium. They are less strongly ionizing than alpha particles and, like cathode rays, are easily deflected by magnetic and electric fields.

boiling point

for any given liquid, the temperature at which the application of heat raises the temperature of the liquid no further, but converts it into vapour.

The boiling point of water under normal pressure is 100°C/212°F. The lower the pressure, the lower the boiling point and vice versa.

Boltzmann constant

constant (symbol k) that relates the kinetic energy (energy of motion) of a gas atom or molecule to temperature. Its value is 1.38066×10^{-23} joules per kelvin. It is equal to the gas constant R, divided by Avogadro's number.

Bose–Einstein condensate

hypothesis put forward in 1925 by Albert Einstein and Indian physicist Satyendra Bose, suggesting that when a dense gas is cooled to a little over absolute zero it will condense and its atoms will lose their individuality and act as an organized whole. The first Bose–Einstein condensate was produced in June 1995 by US physicists cooling rubidium atoms to 10 billionths of a degree above zero. The condensate existed for about a minute before becoming rubidium ice.

boson

elementary particle whose spin can only take values that are whole numbers or zero. Bosons may be classified as gauge bosons (carriers of the four fundamental forces) or mesons. All elementary particles are either bosons or fermions.

Unlike fermions, more than one boson in a system (such as an atom) can possess the same energy state. That is, they do not obey the Pauli exclusion principle. When developed mathematically, this statement is known as the Bose–Einstein law, after its discoverers Indian physicist Satyendra Bose and Albert Einstein.

Boyle's law

law stating that the volume of a given mass of gas at a constant temperature is inversely proportional to its pressure. For example, if the pressure on a gas doubles, its volume will be reduced by a half, and vice versa. The law was discovered in 1662 by Irish physicist and chemist Robert Boyle. See also gas laws.

breeding

process in a reactor in which more fissionable material is produced than is consumed in running the reactor.

For example, plutonium-239 can be made from the relatively plentiful (but nonfissile) uranium-238, or uranium-233 can be produced from thorium. The Pu-239 or U-233 can then be used to fuel other reactors.

bubble chamber

device for observing the nature and movement of atomic particles, and their interaction with radiation. It is a vessel filled with a superheated liquid through which ionizing particles move and collide. The paths of these particles are shown by strings of bubbles, which can be photographed and studied. By using a pressurized liquid medium instead of a gas, it overcomes drawbacks inherent in the earlier cloud chamber. It was invented by US physicist Donald Glaser in 1952. See particle detector.

calorific value

amount of heat generated by a given mass of fuel when it is completely burned. It is measured in joules per kilogram. Calorific values are measured experimentally with a bomb calorimeter.

capacitance, electrical

property of a capacitor that determines how much charge can be stored in it for a given potential difference between its terminals. It is equal to the ratio of the electrical charge stored to the potential difference. The SI unit of capacitance is the farad, but most capacitors have much smaller capacitances, and the microfarad (a millionth of a farad) is the commonly used practical unit.

capacitor or condenser

device for storing electric charge, used in electronic circuits; it consists of two or more metal plates separated by an insulating layer called a dielectric (see capacitance).

Its *capacitance* is the ratio of the charge stored on either plate to the potential difference between the plates.

capillarity

spontaneous movement of liquids up or down narrow tubes, or capillaries. The movement is due to unbalanced molecular attraction at the boundary between the liquid and the tube. If liquid molecules near the boundary are more strongly attracted to molecules in the material of the tube than to other nearby liquid molecules, the liquid will rise in the tube. If liquid molecules are less attracted to the material of the tube than to other liquid molecules, the liquid will fall.

carburation

any process involving chemical combination with carbon, especially the mixing or charging of a gas, such as air, with volatile compounds of carbon (petrol, kerosene, or fuel oil) in order to increase potential heat energy during combustion. Carburation applies to combustion in the cylinders of reciprocating petrol engines of the types used in aircraft, road vehicles, or marine vessels. The device by which the liquid fuel is atomized and mixed with air is called a *carburettor*.

Carnot cycle

series of changes in the physical condition of a gas in a reversible heat engine, necessarily in the following order: (1) isothermal expansion (without change of temperature), (2) adiabatic expansion (without change of heat content), (3) isothermal compression, and (4) adiabatic compression.

The principles derived from a study of this cycle are important in the fundamentals of heat and thermodynamics.

cathode

part of an electronic device in which electrons are generated. In a thermionic valve, electrons are produced by the heating effect of an applied current; in a photocell, they are produced by the interaction of light and a semiconducting material. The cathode is kept at a negative potential relative to the device's other electrodes (anodes) in order to ensure that the liberated electrons stream away from the cathode and towards the anodes.

cathode ray

stream of fast-moving electrons that travel from a cathode (negative electrode) towards an anode (positive electrode) in a vacuum tube. They carry a negative charge and can be deflected by electric and magnetic fields. Cathode rays focused into fine beams of fast electrons are used in cathode-ray tubes, the electrons' kinetic energy being converted into light energy as they collide with the tube's fluorescent screen.

cathode-ray oscilloscope (CRO)

instrument used to measure electrical potentials or voltages that vary over time and to display the waveforms of electrical oscillations or signals. Readings are displayed graphically on the screen of a cathode-ray tube.

The CRO is used as a voltmeter with any voltage change shown on screen by an up (positive) or down (negative) movement of a bright dot. This dot is produced by a beam of electrons hitting the phospor layer on the inside of the screen. Over time the dot traces a graph across the screen showing voltage change against time. If voltage remains constant the graph consists of a horizontal line.

line

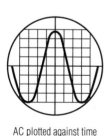 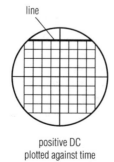 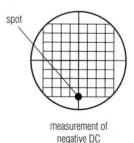

spot

AC plotted against time   positive DC   measurement of
                                 plotted against time   negative DC

The cathode-ray oscilloscope (CRO) is used to measure voltage and its changes over time. This is how the CRO screen would appear measuring AC and DC current.

cathode-ray tube (CRT)

vacuum tube in which a beam of electrons is produced and focused onto a fluorescent screen. The electrons' kinetic energy is converted into light energy as they collide with the screen. It is an essential component of television receivers, computer visual display units, and oscilloscopes.

cation

ion carrying a positive charge. During electrolysis, cations in the electrolyte move to the cathode (negative electrode).

cell, electrical or voltaic cell or galvanic cell

device in which chemical energy is converted into electrical energy; the popular name is 'battery', but this strictly refers to a collection of cells in one unit. The reactive chemicals of a *primary cell* cannot be replenished, whereas *secondary cells* – such as storage batteries – are rechargeable: their chemical reactions can be reversed and the original condition restored by applying an electric current. It is dangerous to attempt to recharge a primary cell.

Each cell contains two conducting electrodes immersed in an electrolyte, in a container. A spontaneous chemical reaction within the cell generates a negative charge (excess of electrons) on one electrode, and a positive charge (deficiency of electrons) on the other. The accumulation of these equal but opposite charges prevents the reaction from continuing unless an outer connection (external circuit) is made between the electrodes allowing the charges to dissipate. When this occurs, electrons escape from the cell's negative terminal and are replaced at the positive, causing a current to flow. After prolonged use, the cell will become flat (ceases to supply current). The first cell was made by Italian physicist Alessandro Volta in 1800. Types of primary cells include the Daniell, Lalande, Leclanché, and so-called 'dry' cells; secondary cells include the Planté, Faure, and Edison. Newer types include the Mallory (mercury depolarizer), which has a very stable discharge curve and can be made in very small units (for example, for hearing aids), and the Venner accumulator, which can be made substantially solid for some purposes. Rechargeable nickel–cadmium dry cells are available for household use.

The reactions that take place in a simple cell depend on the fact that some metals are more reactive than others. If two different metals are joined by an electrolyte and a wire, the more reactive metal loses electrons to form ions. The ions pass into solution in the electrolyte, while the electrons flow down the wire to the less reactive metal. At the less reactive metal the electrons are taken up by the positive ions in the electrolyte, which completes the circuit. If the two metals are zinc and copper and the electrolyte is dilute sulphuric acid, the following cell reactions occur. The zinc atoms dissolve as they lose electrons (oxidation) $Zn - 2e^- \rightarrow Zn^{2+}$. The two electrons travel down the wire and are taken up by the hydrogen ions in the electrolyte (reduction) $2H^+ + 2e^- \rightarrow H_2$. The overall cell reaction is obtained by combining these two reactions; the zinc rod slowly dissolves and bubbles of hydrogen appear at the copper rod

basic principles

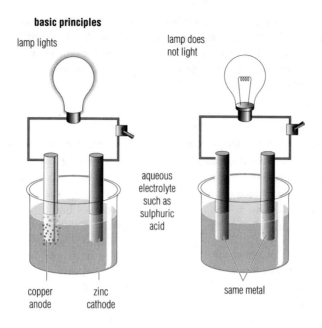

lamp lights

lamp does not light

aqueous electrolyte such as sulphuric acid

copper anode   zinc cathode

same metal

a simple cell

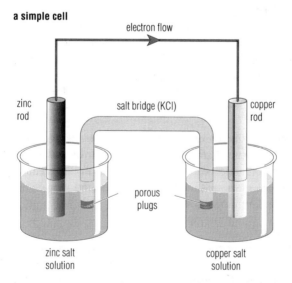

electron flow

zinc rod

salt bridge (KCl)

copper rod

porous plugs

zinc salt solution

copper salt solution

When electrical energy is produced from chemical energy using two metals acting as electrodes in a aqueous solution, it is sometimes known as a galvanic cell or voltaic cell. Here the two metals copper (+) and zinc (–) are immersed in dilute sulphuric acid, which acts as an electrolyte. If a light bulb is connected between the two, an electric current will flow with bubbles of gas being deposited on the electrodes in a process known as polarization.

$Zn + 2H^+ \rightarrow Zn^{2+} + H_2$. If each rod is immersed in an electrolyte containing ions of that metal, and the two electrolytes are joined by a salt bridge, metallic copper deposits on the copper rod as the zinc rod dissolves in a redox reaction, just as if zinc had been added to a copper salt solution $Zn - 2e^- \rightarrow Zn^{2+}$ $Cu^{2+} + 2e^- \rightarrow Cu$ $Zn_{(s)} + Cu^{2+}{}_{(aq)} \rightarrow Zn^{2+}{}_{(aq)} + Cu_{(s)}$.

centre of mass
point in or near an object at which the whole mass of the object may be considered to be concentrated. A symmetrical homogeneous object such as a sphere or cube has its centre of mass at its geometrical centre; a hollow object (such as a cup) may have its centre of mass in space inside the hollow.

For an object to be in stable equilibrium, a vertical line through its centre of mass must run within the boundaries of its base; if tilted until this line falls outside the base, the object becomes unstable and topples over.

centrifugal force
apparent force arising for an observer moving with a rotating system. For an object of mass m moving with a velocity v in a circle of radius r, the centrifugal force F equals mv^2/r (outward).

centripetal force
force that acts radially inward on an object moving in a curved path. For example, with a weight whirled in a circle at the end of a length of string, the centripetal force is the tension in the string. For an object of mass m moving with a velocity v in a circle of radius r, the centripetal force F equals mv^2/r (inward). The reaction to this force is the centrifugal force.

chain reaction
fission reaction that is maintained because neutrons released by the splitting of some atomic nuclei themselves go on to split others, releasing even more neutrons. Such a reaction can be controlled (as in a nuclear reactor) by using moderators to absorb excess neutrons. Uncontrolled, a chain reaction produces a nuclear explosion (as in an atom bomb).

change of state
change in the physical state (solid, liquid, or gas) of a material. For instance, melting, boiling, evaporation, and their opposites, solidification and condensation, are changes of state. The former set of changes are brought about by heating or decreased pressure; the latter by cooling or increased pressure.

These changes involve the absorption or release of heat energy, called latent heat, even though the temperature of the material does not change during the transition between states. See also states of matter. In the unusual change of state called *sublimation,* a solid changes directly to a gas without passing through the liquid state. For example, solid carbon dioxide (dry ice) sublimes to carbon dioxide gas.

charge
see electric charge.

charm
property possessed by one type of quark (very small particles found inside protons and neutrons), called the charm quark. The effects of charm are only seen in experiments with particle accelerators. See elementary particles.

Cherenkov radiation
type of electromagnetic radiation emitted by charged particles entering a transparent medium at a speed greater than the speed of light in the medium. It appears as a bluish light. Cherenkov radiation can be detected from high-energy cosmic rays entering the Earth's atmosphere. It is named after Pavel Alexseevich Cherenkov, the Russian physicist who first observed it.

choke coil
coil employed to limit or suppress alternating current without stopping direct current, particularly the type used as a 'starter' in the circuit of fluorescent lighting.

circuit
arrangement of electrical components through which a current can flow. There are two basic circuits, series and parallel. In a series circuit, the components are connected end to end so that the current flows through all components one after the other. In a parallel circuit, components are connected side by side so that part of the current passes through each component. A circuit diagram shows in graphical form how components are connected together, using standard symbols for the components.

cloud chamber
apparatus, now obsolete, for tracking ionized particles. It consists of a vessel fitted with a piston and filled with air or other gas, saturated with water vapour. When the volume of the vessel is suddenly expanded by moving the piston outwards, the vapour cools and a cloud of tiny droplets forms on any nuclei, dust, or ions present. As fast-moving ionizing particles collide with the air or gas molecules, they show as visible tracks.

Much information about interactions between such particles and radiations has been obtained from photographs of these tracks. The system was improved upon by the use of liquid hydrogen or helium instead of air or gas (see particle detector) and by the spark chamber. The cloud chamber was devised in 1897 by Charles Thomson Rees Wilson (1869–1959) at Cambridge University.

cold fusion
fusion of atomic nuclei at room temperature. If cold fusion were possible it would provide a limitless, cheap, and pollution-free source of energy, and it has therefore been the subject of research around the world.

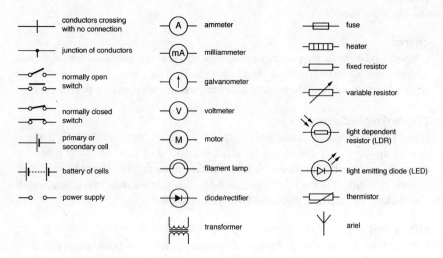

Electrical symbols commonly used in circuit diagrams.

In 1989, Martin Fleischmann and Stanley Pons of the University of Utah, USA, claimed that they had achieved cold fusion in the laboratory, but their results could not be substantiated. The University of Utah announced in 1998 that they would allow the cold fusion patent to elapse, given that the work of Pons and Fleischmann has never been reproduced.

colour
quality or wavelength of light emitted or reflected from an object. Visible white light consists of electromagnetic radiation of various wavelengths, and if a beam is refracted through a prism, it can be spread out into a spectrum, in which the various colours correspond to different wavelengths. From long to short wavelengths (from about 700 to 400 nanometres) the colours, traditionally given following Newton, are red, orange, yellow, green, blue, indigo, and violet.

The light entering our eyes is either emitted by hot or luminous objects or reflected from the objects we see.

emitted light
Sources of light have a characteristic spectrum or range of wavelengths. Hot solid objects emit light with a broad range of wavelengths, the maximum intensity being at a wavelength that depends on the temperature. The hotter the object, the shorter the wavelengths emitted, as described by Wien's displacement law. Low-pressure gases, such as the vapour of sodium street lights, can emit light at discrete wavelengths. The pattern of wavelengths emitted is unique to each gas and can be used to identify the gas (see spectroscopy).

reflected light
When an object is illuminated by white light, some of the wavelengths are absorbed and some are reflected to the eye of an observer. The object appears

coloured because of the mixture of wavelengths in the reflected light. For instance, a red object absorbs all wavelengths falling on it except those in the red end of the spectrum. This process of subtraction also explains why certain mixtures of paints produce different colours. Blue and yellow paints when mixed together produce green because between them the yellow and blue pigments absorb all wavelengths except those around green. A suitable combination of three pigments – cyan (blue-green), magenta (blue-red), and yellow – can produce any colour when mixed.

primary colours
In the light-sensitive lining of our eyeball (the retina), cells called cones are responsible for colour vision. There are three kinds of cones. Each type is sensitive to one colour only, either red, green, or blue. The brain combines the signals sent from the set of cones to produce a sensation of colour. When all cones are stimulated equally the sensation is of white light. The three colours to which the cones respond are called the *additive primary colours*. By mixing lights of these three colours, it is possible to produce any colour. This process is called colour mixing by addition, and is used to produce the colour on a television screen, the inside of which is coated with glowing phosphor dots of red, green, and blue.

complementary colours
Pairs of colours that produce white light, such as yellow and blue, are called complementary colours.

classifying colours
Many schemes have been proposed for classifying colours. The most widely used is the Munsell scheme, which classifies colours according to their hue (dominant wavelength), saturation (the degree of whiteness), and lightness (intensity).

colour
in elementary particle physics, a property of quarks analogous to electric charge but having three states, denoted by red, green, and blue. The colour of a quark is changed when it emits or absorbs a gluon. The term has nothing to do with colour in its usual sense. See quantum chromodynamics.

Compton effect
increase in wavelength (loss of energy) of a photon by its collision with a free electron (*Compton scattering*). The Compton effect was first demonstrated with X-rays and provided early evidence that electromagnetic waves consisted of particles – photons – that carried both energy and momentum. It is named after US physicist Arthur Compton.

condenser
former name for a *capacitor*.

conduction, electrical

flow of charged particles through a material giving rise to electric current. Conduction in metals involves the flow of negatively charged free electrons. Conduction in gases and some liquids involves the flow of ions that carry positive charges in one direction and negative charges in the other. Conduction in a semiconductor such as silicon involves the flow of electrons and positive holes.

Conventionally, current is regarded as a movement of positive electricity from points at high potential to points at a lower potential. Metals contain many free electrons that move with little hindrance from the parent atoms, and they are therefore good conductors. In other solids the electrons are more tightly bound to the atoms and conduction is less. Increase of temperature in this case frees more electrons, so the conductivity of nonmetals increases with rising temperature. Conduction in gases and in many liquids involves a flow not merely of electrons, but of atoms or groups of atoms as well. When a salt such as sodium chloride is dissolved in water, the chlorine atoms each gain an electron and become negatively charged, while the sodium atoms each lose one and become positively charged. These charged atoms, or ions, can move through the liquid and transport electricity (electrolysis). Gases are, under normal circumstances, almost completely nonconducting. They may be ionized by irradiation with X-rays or by radioactive radiations. They are more readily maintained in a conducting state at high temperatures, as in the electric arc, or at low pressures, as in electric discharge lamps. At very low temperatures certain metals such as lead become almost perfect conductors, and if a current is set up in a ring of a metal in this superconducting state, the current persists for a long time without any energy being supplied. In all other cases the flow of a current through a conductor is accompanied by a loss of energy as heat, and a continuous supply of energy is required to maintain the current. A magnetic field (see magnetism) is always present in the space around a conductor in which a current is flowing.

conduction, heat

flow of heat energy through a material without the movement of any part of the material itself (compare conduction, electrical). Heat energy is present in all materials in the form of the kinetic energy of their vibrating molecules, and may be conducted from one molecule to the next in the form of this mechanical vibration. In the case of metals, which are particularly good conductors of heat, the free electrons within the material transport heat around very quickly.

conservation of momentum

law of mechanics that states that total momentum is conserved (remains constant) in all collisions, providing no external resultant force acts on the colliding bodies. The principle may be expressed as an equation used to solve numerical problems: total momentum before collision = total momentum after collision.

cosmic radiation

streams of high-energy particles and elctromagnetic radiation from outer space, consisting of electrons, protons, alpha particles, light nuclei, and gamma rays, which collide with atomic nuclei in the Earth's atmosphere, and produce secondary nuclear particles (chiefly mesons, such as pions and muons) that shower the Earth.

Those of lower energy seem to be galactic in origin, and those of high energy of extragalactic origin. The galactic particles may come from supernova explosions or pulsars. At higher energies, other sources must exist, possibly the giants jets of gas that are emitted from some galaxies.

coulomb

SI unit (symbol C) of electrical charge. One coulomb is the quantity of electricity conveyed by a current of one ampere in one second. The unit is named after French scientist Charles Coulomb (1736–1806).

critical mass

minimum mass of fissile material that can undergo a continuous chain reaction. Below this mass, too many neutrons escape from the surface for a chain reaction to carry on; above the critical mass, the reaction may accelerate into a nuclear explosion.

cryogenics

science of very low temperatures (approaching absolute zero), including the production of very low temperatures and the exploitation of special properties associated with them, such as the disappearance of electrical resistance (superconductivity).

Low temperatures can be produced by the Joule–Thomson effect (cooling a gas by making it do work as it expands). Gases such as oxygen, hydrogen, and helium may be liquefied in this way, and temperatures of 0.3K can be reached. Further cooling requires magnetic methods; a magnetic material, in contact with the substance to be cooled and with liquid helium, is magnetized by a strong magnetic field. The heat generated by the process is carried away by the helium. When the material is then demagnetized, its temperature falls; temperatures of around 10^{-3}K have been achieved in this way. Much lower temperatures, of a few billionths of a kelvin above absolute zero, have been obtained by trapping gas atoms in an 'optical molasses' of crossed laser beams, which slow them down. At temperatures near absolute zero, materials can display unusual properties. Some metals, such as mercury and lead, exhibit superconductivity. Liquid helium loses its viscosity and becomes a 'superfluid' when cooled to below 2K; in this state it flows up the sides of its container. Cryogenics has several practical applications. Cryotherapy is a process used in eye surgery, in which a freezing probe is briefly applied to repair a break in the retina. Electronic components called Josephson junctions, which could be used in very fast computers, need low temperatures to function. Magnetic levitation (maglev) systems must be maintained at low temperatures.

current, electric

see electric current.

cycle

sequence of changes that moves a system away from, and then back to, its original state. An example is a vibration that moves a particle first in one direction and then in the opposite direction, with the particle returning to its original position at the end of the vibration.

cyclotron

circular type of particle accelerator.

DC

abbreviation for *direct current* (electricity).

decay, radioactive

see radioactive decay.

decibel

unit (symbol dB) of measure used originally to compare sound intensities and subsequently electrical or electronic power outputs; now also used to compare voltages. An increase of 10 dB is equivalent to a 10-fold increase in intensity or power. The decibel scale is used for audibility measurements, as one decibel, representing an increase of about 25%, is about the smallest change the human ear can detect. A whisper has an intensity of 20 dB; 140 dB (a jet aircraft taking off nearby) is the threshold of pain.

The difference in decibels between two levels of intensity (or power) L_1 and L_2 is 10 $\log_{10}(L_1/L_2)$; a difference of 1 dB thus corresponds to a level of $10^{0.1}$, which is about1.026.

density

measure of the compactness of a substance; it is equal to its mass per unit volume and is measured in kg per cubic metre/lb per cubic foot. Density is a scalar quantity. The average density D of a mass m occupying a volume V is given by the formula:

$$D = \frac{m}{V}$$

Relative density is the ratio of the density of a substance to that of water at 4°C/39.2°F.

In electricity, current density is the amount of current passing through a cross-sectional area of a conductor (usually given in amperes per sq in or per sq cm).

diffraction

spreading out of waves when they pass through a small gap or around a small object, resulting in some change in the direction of the waves. In order for this effect to be observed the size of the object or gap must be comparable to or smaller than the wavelength of the waves. Diffraction occurs with all forms of progressive waves – electromagnetic, sound, and water waves – and explains such phenomena as why long-wave radio waves can bend round hills better than short-wave radio waves.

The wavelength of light ranges from 400 nm to about 700 nm, a few orders of magnitude smaller than radio waves. The gap through which light travels must be extremely small to observe diffraction. The slight spreading of a light beam through a narrow slit causes the different wavelengths of light to interfere with each other to produce a pattern of light and dark bands. A *diffraction grating* is a plate of glass or metal ruled with close, equidistant parallel lines used for separating a wave train such as a beam of incident light into its component frequencies (white light results in a spectrum). The wavelength of sound is between 0.5 m/1.6 ft and 2.0 m/6.6 ft. When sound waves travel through doorways or between buildings they are diffracted significantly, so that the sound is heard round corners.

The regularly spaced atoms in crystals diffract X-rays, and by exploiting this fact the structure of many substances has been elucidated, including that of proteins.

dip, magnetic

angle at a particular point on the Earth's surface between the direction of the Earth's magnetic field and the horizontal. It is measured using a *dip circle*, which has a magnetized needle suspended so that it can turn freely in the vertical plane of the magnetic field. In the northern hemisphere the needle dips below the horizontal, pointing along the line of the magnetic field towards its north pole. At the magnetic north and south poles, the needle dips vertically and the angle of dip is 90°.

dipole

uneven distribution of magnetic or electrical characteristics within a molecule or substance so that it behaves as though it possesses two equal but opposite poles or charges, a finite distance apart.

The uneven distribution of electrons within a molecule composed of atoms of different electronegativities may result in an apparent concentration of electrons towards one end of the molecule and a deficiency towards the other, so that it forms a dipole consisting of apparently separated but equal positive and negative charges. The product of one charge and the distance between them is the *dipole moment*. A bar magnet has a magnetic dipole and behaves as though its magnetism were concentrated in separate north and south magnetic poles.

direct current (DC)

electric current that flows in one direction, and does not reverse its flow as alternating current does. The electricity produced by a battery is direct current.

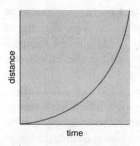 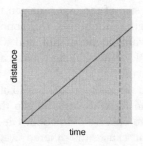 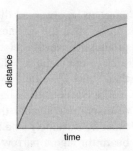

In a distance-time graph distance travelled by an object is plotted against the time it has taken for the object to travel that distance. If the line plotted is straight then the object is travelling at a uniform rate. The line curving upward shows acceleration; a curve downwards plots deceleration.

distance ratio

in a machine, the distance moved by the input force, or effort, divided by the distance moved by the output force, or load. The ratio indicates the movement magnification achieved, and is equivalent to the machine's velocity ratio.

distance-time graph

graph used to describe the motion of a body by illustrating the relationship between the distance that it travels and the time taken. Plotting distance (on the vertical axis) against time (on the horizontal axis) produces a graph the gradient of which is the body's speed. If the gradient is constant (the graph is a straight line), the body has uniform or constant speed; if the gradient varies (the graph is curved), then so does the speed and the body may be said to be accelerating or decelerating.

domain

small area in a magnetic material that behaves like a tiny magnet. The magnetism of the material is due to the movement of electrons in the atoms of the domain. In an unmagnetized sample of material, the domains point in random directions, or form closed loops, so that there is no overall magneti-zation of the sample. In a magnetized sample, the domains are aligned so that their magnetic effects combine to produce a strong overall magnetism.

Doppler effect

change in the observed frequency (or wavelength) of waves due to relative motion between the wave source and the observer. The Doppler effect is responsible for the perceived change in pitch of a siren as it approaches and then recedes, and for the red shift of light from distant galaxies. It is named after the Austrian physicist Christian Doppler.

dynamics or kinetics

mathematical and physical study of the behaviour of bodies under the action of forces that produce changes of motion in them.

dynamo

simple generator or machine for transforming mechanical energy into electrical energy. A dynamo in basic form consists of a powerful field magnet between the poles of which a suitable conductor, usually in the form of a coil (armature), is rotated. The mechanical energy of rotation is thus converted into an electric current in the armature.

Present-day dynamos work on the principles described by English physicist Michael Faraday in 1830, that an electromotive force is developed in a conductor when it is moved in a magnetic field. The dynamo that powers the lights on a bicycle is an example of an alternator, that is, it produces alternating current (AC).

echo

repetition of a sound wave, or of a radar or sonar signal, by reflection from a surface. By accurately measuring the time taken for an echo to return to the transmitter, and by knowing the speed of a radar signal (the speed of light) or a sonar signal (the speed of sound in water), it is possible to calculate the range of the object causing the echo (echolocation).

A similar technique is used in echo sounders to estimate the depth of water under a ship's keel or the depth of a shoal of fish.

eddy current

electric current induced, in accordance with Faraday's laws of electromagnetic induction, in a conductor located in a changing magnetic field. Eddy currents can cause much wasted energy in the cores of transformers and other electrical machines.

efficiency

general term indicating the degree to which a process or device can convert energy from one form to another without loss. It is normally expressed as a fraction or percentage, where 100% indicates conversion with no loss. The efficiency of a machine, for example, is the ratio of the work done by the machine to the energy put into the machine; in practice it is always less than 100% because of frictional heat losses. Certain electrical machines with no moving parts, such as transformers, can approach 100% efficiency.

Since the mechanical advantage, or force ratio, is the ratio of the load (the output force) to the effort (the input force), and the velocity ratio is the distance moved by the effort divided by the distance moved by the load, for certain machines the efficiency can also be defined as the mechanical advantage divided by the velocity ratio.

In the special case of a *heat engine*, the efficiency can never exceed $1 - T_2/T_1$, where T_1 is the absolute temperature of the heat source and T_2 is the absolute temperature of the exhaust.

elasticity

ability of a solid to recover its shape once deforming forces (stresses modifying its dimensions or shape) are removed. An elastic material obeys Hooke's law, which states that its deformation is proportional to the applied stress up to a certain point, called the *elastic limit*, beyond which additional stress will deform it permanently. Elastic materials include metals and rubber; however, all materials have some degree of elasticity.

E layer formerly Kennelly–Heaviside layer

the lower regions (90–120 km/56–75 mi) of the ionosphere, which reflect radio waves, allowing their reception around the surface of the Earth. The E layer approaches the Earth by day and recedes from it at night.

electric charge

property of some bodies that causes them to exert forces on each other. Two bodies both with positive or both with negative charges repel each other, whereas bodies with opposite or 'unlike' charges attract each other. Electrons possess a negative charge, and protons an equal positive charge. The SI unit of electric charge is the coulomb (symbol C).

A body can be charged by friction, induction, or chemical change and shows itself as an accumulation of electrons (negative charge) or loss of electrons (positive charge) on an atom or body. Atoms generally have zero net charge but can gain electrons to become negative ions or lose them to become positive ions. So-called static electricity, seen in such phenomena as the charging of nylon shirts when they are pulled on or off, or in brushing hair, is in fact the gain or loss of electrons from the surface atoms. A flow of charge (such as electrons through a copper wire) constitutes an electric current; the flow of current is measured in amperes (symbol A).

electric current

flow of electrically charged particles through a conducting circuit due to the presence of a potential difference. The current at any point in a circuit is the amount of charge flowing per second; its SI unit is the ampere (coulomb per second).

Current carries electrical energy from a power supply, such as a battery of electrical cells, to the components of the circuit, where it is converted into other forms of energy, such as heat, light, or motion. It may be either direct current or alternating current.

heating effect

When current flows in a component possessing resistance, electrical energy is converted into heat energy. If the resistance of the component is R ohms and the current through it is I amperes, then the heat energy W (in joules) generated in a time t seconds is given by the formula:

$$W = I^2Rt$$

or power P (in watts) is given by:

$$P = I^2R$$

magnetic effect

A magnetic field is created around all conductors that carry a current. When a current-bearing conductor is made into a coil it forms an electromagnet with a magnetic field that is similar to that of a bar magnet, but which disappears as soon as the current is switched off. The strength of the magnetic field is directly proportional to the current in the conductor – a property that allows a small electromagnet to be used to produce a pattern of magnetism on recording tape that accurately represents the sound or data stored. The direction of the field created around a conducting wire may be predicted by using Maxwell's screw rule.

motor effect

A conductor carrying current in a magnetic field experiences a force, and is impelled to move in a direction perpendicular to both the direction of the current and the direction of the magnetic field. The direction of motion may be predicted by Fleming's left-hand rule (see Fleming's rules). The magnitude of the force experienced depends on the length of the conductor and on the strengths of the current and the magnetic field, and is greatest when the conductor is at right angles to the field. A conductor wound into a coil that can rotate between the poles of a magnet forms the basis of an electric motor.

electric field

region in which a particle possessing electric charge experiences a force owing to the presence of another electric charge. The strength of an electric field, E, is measured in volts per metre ($V\ m^{-1}$). It is a type of electromagnetic field.

electricity

all phenomena caused by electric charge, whether static or in motion. Electric charge is caused by an excess or deficit of electrons in the charged substance, and an electric current is the movement of charge through a material. Substances may be electrical conductors, such as metals, that allow the passage of electricity through them readily, or insulators, such as rubber, that are extremely poor conductors. Substances with relatively poor conductivities that increase with a rise in temperature or when light falls on the material, are known as semiconductors.

electrical properties of solids

The first artificial electrical phenomenon to be observed was that some naturally occurring materials such as amber, when rubbed with a piece of cloth, would then attract small objects such as dust and pieces of paper. Rubbing the object caused it to become electrically charged so that it had an excess or deficit of electrons. When the amber is rubbed with a piece of cloth electrons are transferred from the cloth to the amber so that the amber has an excess of electrons and is negatively charged, and the cloth has a deficit of electrons and is positively charged. This accumulation of charge is called static electricity. This charge on the object exerts an electric field in the space around itself that can attract or repel other objects. It was discovered that there are only two types of charge, positive and negative, and that they neutralize one another. Objects with a like charge always repel one another while objects with an unlike charge attract each other. Neutral objects (such as pieces of paper) can be attracted to charged bodies by electrical induction. For example, the charge on a negatively charged body causes a separation of charge across the neutral body. The positive charges tend to move towards the side near the negatively charged body and the negative charges tend to move towards the opposite side so that the neutral body is weakly attracted to the charged body by induction.

current, charge, and energy

An electric current in a material is the passage of charge through it. In metals and other conducting materials, the charge is carried by free electrons that are not bound tightly to the atoms and are thus able to move through the material. For charge to flow in a circuit there must be a potential difference (pd) applied across the circuit. This is often supplied in the form of a battery that has a positive terminal and a negative terminal. Under the influence of the potential difference, the electrons are repelled from the negative terminal side of the circuit and attracted to the positive terminal of the battery. A steady flow of electrons around the circuit is produced. Current flowing through a circuit can be measured using an ammeter and is measured in amperes (or amps). A coulomb (C) is the unit of charge, defined as the charge passing a point in a wire each second when the current is exactly 1 amp. The unit of charge is named after Charles Augustin de Coulomb. Direct current (DC) flows continuously in one direction; alternating current (AC) flows alternately in each direction. In a circuit the battery provides energy to make charge flow through the circuit. The amount of energy supplied to each unit of charge is called the electromotive force (emf). The unit of emf is the volt (V). A battery has an emf of 1 volt when it supplies 1 joule of energy to each coulomb of charge flowing through it. The energy carried by flowing charges can be used to do work, for example to light a bulb, to cause current to flow through a resistor, to emit radiation, or to produce heat. When the energy carried by a current is made to do work in this way, a potential difference can be measured across the circuit component concerned by a voltmeter or a cathode-ray oscilloscope. The potential difference is also measured in volts. Power, measured in watts, is the product of current and voltage. Potential difference and current measure are related to one another. This relationship was discovered by Georg Ohm, and is expressed by Ohm's law: the current through a wire is proportional to the potential difference across its ends. The potential difference divided by the current is a constant for a given piece of wire. This constant for a given material is called the resistance.

conduction in liquids and gases

In liquids, current can flow by the movement of charged ions through a solution or molten salt (the electrolyte), resulting in the migration of ions to the electrodes: positive ions (cations) to the negative electrode (cathode) and negative ions (anions) to the positive electrode (anode). This process is called electrolysis and represents bi-directional flow of charge as opposite charges move to oppositely charged electrodes. In metals, charges are only carried by free electrons and therefore move in only one direction.

electromagnetism
Magnetic fields are produced either by current-carrying conductors or by permanent magnets. In current-carrying wires, the magnetic field lines are concentric circles around the wire. Their direction depends on the direction of the current and their strength on the size of the current. If a conducting wire is moved within a magnetic field, the magnetic field acts on the free electrons within the conductor, displacing them and causing a current to flow. The force acting on the electrons and causing them to move is greatest when the wire is perpendicular to the magnetic field lines. The direction of the current is given by the left-hand rule. The generation of a current by the relative movement of a conductor in a magnetic field is called electromagnetic induction. This is the basis of how a dynamo works.

electrode
any terminal by which an electric current passes in or out of a conducting substance; for example, the anode or cathode in a battery or the carbons in an arc lamp. The terminals that emit and collect the flow of electrons in thermionic valves (electron tubes) are also called electrodes: for example, cathodes, plates, and grids.

electrodynamics
branch of physics dealing with electric charges, electric currents and associated forces. Quantum electrodynamics (QED) studies the interaction between charged particles and their emission and absorption of electromagnetic radiation. This subject combines quantum theory and relativity theory, making accurate predictions about subatomic processes involving charged particles such as electrons and protons.

electromagnet
coil of wire wound around a soft iron core that acts as a magnet when an electric current flows through the wire. Electromagnets have many uses: in switches, electric bells, solenoids, and metal-lifting cranes.

The strength of the electromagnet can be increased by increasing the current through the wire, changing the material of the core, or by increasing the number of turns in the wire coil. At the north pole of the electromagnet current flows anticlockwise; at the south pole flow is clockwise.

electromagnetic field
region in which a particle with an electric charge experiences a force. If it does so only when moving, it is in a pure *magnetic field*; if it does so when stationary, it is in an *electric field*. Both can be present simultaneously.

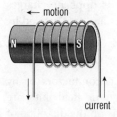 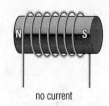 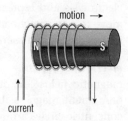

← motion   motion →

current   no current   current

Movement of a magnet in a coil of wire induces a current.

electromagnetic force

one of the four fundamental forces of nature, the other three being the gravitational force, the weak nuclear force, and the strong nuclear force. The particle that is the carrier for the electromagnetic force is the photon.

electromagnetic induction

production of an electromotive force (emf) in a circuit by a change of magnetic flux through the circuit or by relative motion of the circuit and the magnetic flux. In a closed circuit an induced current will be produced. All dynamos and generators make use of this effect. When magnetic tape is driven past the playback head (a small coil) of a tape-recorder, the moving magnetic field induces an emf in the head, which is then amplified to reproduce the recorded sounds.

If the change of magnetic flux is due to a variation in the current flowing in the same circuit, the phenomenon is known as self-induction; if it is due to a change of current flowing in another circuit it is known as mutual induction.

Lenz's law

The direction of an electromagnetically induced current (generated by moving a magnet near a wire or a wire in a magnetic field) will be such as to oppose the motion producing it. This law is named after the German physicist Heinrich Friedrich Lenz (1804–1865), who announced it in 1833.

Faraday's laws

English scientist Michael Faraday proposed three laws of electromagnetic induction: (1) a changing magnetic field induces an electromagnetic force in a conductor; (2) the electromagnetic force is proportional to the rate of change of the field; (3) the direction of the induced electromagnetic force depends on the orientation of the field.

electromagnetic waves

oscillating electric and magnetic fields travelling together through space at a speed of nearly 300,000 km/186,000 mi per second. The (limitless) range of possible wavelengths and frequencies of electromagnetic waves, which can be thought of as making up the *electromagnetic spectrum*, includes radio waves, infrared radiation, visible light, ultraviolet radiation, X-rays, and gamma rays.

Radio and television waves lie at the *long wavelength–low frequency* end of the spectrum, with wavelengths longer than 10 $^{-4}$ m. Infrared radiation has wavelengths between 10 $^{-4}$ m and 7 × 10^{-7} m. Visible light has yet shorter

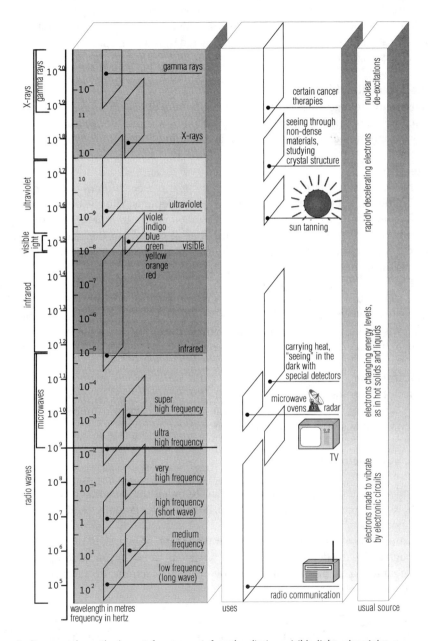

Radio waves have the lowest frequency. Infrared radiation, visible light, ultraviolet radiation, X-rays, and gamma rays have progressively higher frequencies.

181

wavelengths from 7×10^{-7} m to 4×10^{-7} m. Ultraviolet radiation is near the *short wavelength–high frequency* end of the spectrum, with wavelengths between 4×10^{-7} m and 10^{-8} m. X-rays have wavelengths from 10^{-8} m to 10^{-12} m. Gamma radiation has the shortest wavelengths of less than 10^{-10} m. The different wavelengths and frequencies lend specific properties to electromagnetic waves. While visible light is diffracted by a diffraction grating, X-rays can only be diffracted by crystals. Radio waves are refracted by the atmosphere; visible light is refracted by glass or water.

electromotive force (emf)

loosely, the voltage produced by an electric battery or generator in an electrical circuit or, more precisely, the energy supplied by a source of electric power in driving a unit charge around the circuit. The unit is the volt.

electron

stable, negatively charged elementary particle; it is a constituent of all atoms, and a member of the class of particles known as leptons. The electrons in each atom surround the nucleus in groupings called shells; in a neutral atom the number of electrons is equal to the number of protons in the nucleus. This electron structure is responsible for the chemical properties of the atom (see atomic structure).

Electrons carry a charge of 1.602177×10^{-19} coulomb and have a mass of 9.109×10^{-31} kg, which is 1/1836 times the mass of a proton. A beam of electrons will undergo diffraction (scattering) and produce interference patterns in the same way as electromagnetic waves such as light; hence they may be regarded as waves as well as particles.

electron gun

series of electrodes, including a cathode for producing an electron beam. It plays an essential role in many electronic devices, including cathode-ray tubes (television tubes) and electron microscopes.

electrons, delocalized

electrons that are not associated with individual atoms or identifiable chemical bonds, but are shared collectively by all the constituent atoms or ions of some chemical substances (such as metals, graphite, and aromatic compounds).

A metallic solid consists of a three-dimensional arrangement of metal ions through which the delocalized electrons are free to travel. Aromatic compounds are characterized by the sharing of delocalized electrons by several atoms within the molecule.

electroscope

apparatus for detecting electric charge. The simple gold-leaf electroscope consists of a vertical conducting (metal) rod ending in a pair of rectangular pieces of gold foil, mounted inside and insulated from an earthed metal case or glass jar. An electric charge applied to the end of the metal rod makes the gold leaves diverge, because they each receive a similar charge (positive or negative) and so repel each other.

The polarity of the charge can be found by bringing up another charge of known polarity and applying it to the metal rod. A like charge has no effect on the gold leaves, whereas an opposite charge neutralizes the charge on the leaves and causes them to collapse.

electrostatics
study of stationary electric charges and their fields (not currents). See static electricity.

elementary particle
subatomic particle that is not known to be made up of smaller particles, and so can be considered one of the fundamental units of matter. There are three groups of elementary particles: quarks, leptons, and gauge bosons.

Quarks, of which there are 12 types (up, down, charm, strange, top, and bottom, plus the antiparticles of each), combine in groups of three to produce heavy particles called baryons, and in groups of two to produce intermediate-mass particles called mesons. They and their composite particles are influenced by the strong nuclear force. *Leptons* are particles that do not interact via the strong nuclear force. Again, there are 12 types: the electron, muon, tau; their neutrinos, the electron neutrino, muon neutrino, and tau neutrino; and the antiparticles of each. These particles are influenced by the weak nuclear force, as well as by gravitation and electromagnetism. *Gauge bosons* carry forces between other particles. There are four types: gluon, photon, intermediate vector bosons (W+, W-, and Z), as yet unobserved, and graviton. The gluon carries the strong nuclear force, the photon the electromagnetic force, W+, W-, and Z the weak nuclear force, and the graviton the force of gravity (see forces, fundamental).

emf
abbreviation for *electromotive force.*

energy
capacity for doing work. Energy can exist in many different forms. For example, potential energy (PE) is energy deriving from position; thus a stretched spring has elastic PE, and an object raised to a height above the Earth's surface, or the water in an elevated reservoir, has gravitational PE. Moving bodies possess kinetic energy (KE). Energy can be converted from one form to another, but the total quantity in a system stays the same (in accordance with the conservation of energy principle). Energy cannot be created or destroyed. For example, as an apple falls it loses gravitational PE but gains KE. Although energy is never lost, after a number of conversions it tends to finish up as the kinetic energy of random motion of molecules (of the air, for example) at relatively low temperatures. This is 'degraded' energy that is difficult to convert back to other forms.

energy types and transfer

A flat battery in a torch will not light the torch. If the battery is fully charged, it contains enough chemical energy to illuminate the torch bulb. When one body A does work on another body B, A transfers energy to B. The energy transferred is equal to the work done by A on B. Energy is therefore measured in joules. The rate of doing work or consuming energy is called power and is measured in watts (joules per second). Energy can be converted from any form into another. A ball resting on a slope possesses potential energy that is gradually changed into kinetic energy of rotation and translation as the ball rolls down. As a pendulum swings, energy is constantly being changed from a potential form at the highest points of the swing to kinetic energy at the lowest point. At positions in between these two extremes, the system possesses both kinetic and potential energy in varying proportions. A weightlifter changes chemical energy from the muscles into potential energy of the weight when the weight is lifted. If the weightlifter releases the weight, the potential energy is converted to kinetic energy as it falls, and this in turn is converted to heat energy and sound energy as it hits the floor. A lump of coal and a tank of petrol, together with the oxygen needed for their combustion, have chemical energy. Other sorts of energy include electrical and nuclear energy, and light and sound. However, all of these types are ultimately classifiable as either kinetic or potential energy.

heat transfer

A difference in temperature between two objects in thermal contact leads to the transfer of energy as heat. Heat is energy transferred due to a temperature difference. Heat is transferred by the movement of particles (that possess kinetic energy) by conduction, convection, and radiation. Conduction involves the movement of heat through a solid material by the movement of free electrons. For example, thermal energy is lost from a house by conduction through the walls and windows. Convection involves the transfer of energy by the movement of fluid particles. All objects radiate heat in the form of radiation of electromagnetic waves. Hotter objects emit more energy than cooler objects. Methods of reducing energy transfer as heat through the use of insulation are important because the world's fuel reserves are limited and heating homes costs a lot of money in fuel bills. Heat transfer from the home can be reduced by a variety of methods, such as loft insulation, cavity wall insulation, and double glazing. The efficiencies of insulating materials in the building industry are compared by measuring their heat-conducting properties, represented by a U-value. A low U-value indicates a good insulating material.

$E = mc^2$

It is now recognized that mass can be converted into energy under certain conditions, according to Einstein's theory of relativity. This conversion of mass into energy is the basis of atomic power. Einstein's special theory of relativity (1905) correlates any gain, E, in energy with a gain, m, in mass, by the equation $E = mc^2$, in which c is the speed of light. The conversion of mass into energy in accordance with this equation applies universally, although it is only for nuclear reactions that the percentage change in mass is large enough to detect.

engine

device for converting stored energy into useful work or movement. Most engines use a fuel as their energy store. The fuel is burnt to produce heat energy – hence the name 'heat engine' – which is then converted into movement. Heat engines can be classified according to the fuel they use (petrol engine or diesel engine), or according to whether the fuel is burnt inside (internal combustion engine) or outside (steam engine) the engine, or according to whether they produce a reciprocating or a rotary motion (turbine or Wankel engine).

entropy

parameter representing the state of disorder of a system at the atomic, ionic, or molecular level; the greater the disorder, the higher the entropy. Thus the fast-moving disordered molecules of water vapour have higher entropy than those of more ordered liquid water, which in turn have more entropy than the molecules in solid crystalline ice.

In a closed system undergoing change, entropy is a measure of the amount of energy unavailable for useful work. At absolute zero ($-273.15°C/-459.67°F/$ 0 K), when all molecular motion ceases and order is assumed to be complete, entropy is zero.

equilibrium

unchanging condition in which an undisturbed system can remain indefinitely in a state of balance. In a *static equilibrium*, such as an object resting on the floor, there is no motion. In a *dynamic equilibrium*, in contrast, a steady state is maintained by constant, though opposing, changes. For example, in a sealed bottle half-full of water, the constancy of the water level is a result of molecules evaporating from the surface and condensing on to it at the same rate.

evaporation

process in which a liquid turns to a vapour without its temperature reaching boiling point. A liquid left to stand in a saucer eventually evaporates because, at any time, a proportion of its molecules will be fast enough (have enough kinetic energy) to escape through the attractive intermolecular forces at the liquid surface into the atmosphere. The temperature of the liquid tends to fall because the evaporating molecules remove energy from the liquid. The rate of evaporation rises with increased temperature because as the mean kinetic energy of the liquid's molecules rises, so will the number possessing enough energy to escape.

A fall in the temperature of the liquid, known as the *cooling effect*, accompanies evaporation because as the faster molecules escape through the surface the mean energy of the remaining molecules falls. The effect may be noticed when wet clothes are worn, and sweat evaporates from the skin. In the body it plays a part in temperature control.

The evaporation of liquid water to form water vapour is responsible for the movement of water from the Earth's surface to the atmosphere.

exclusion principle

principle of atomic structure originated by Austrian–born US physicist Wolfgang Pauli. It states that no two electrons in a single atom may have the same set of quantum numbers.

Hence, it is impossible to pack together certain elementary particles, such as electrons, beyond a certain critical density, otherwise they would share the same location and quantum number. A white dwarf star, which consists of electrons and other elementary particles, is thus prevented from contracting further by the exclusion principle and never collapses. Elementary particles in the class fermions obey the exclusion principle whilst those in the class bosons do not.

expansion

increase in size of a constant mass of substance caused by, for example, increasing its temperature (thermal expansion) or its internal pressure. The *expansivity*, or coefficient of thermal expansion, of a material is its expansion (per unit volume, area, or length) per degree rise in temperature.

farad

SI unit (symbol F) of electrical capacitance (how much electric charge a capacitor can store for a given voltage). One farad is a capacitance of one coulomb per volt. For practical purposes the microfarad (one millionth of a farad, symbol μF) is more commonly used.

The farad is named after English scientist Michael Faraday.

Faraday's laws

three laws of electromagnetic induction, and two laws of electrolysis, all proposed originally by English scientist Michael Faraday: *induction* (1) a changing magnetic field induces an electromagnetic force in a conductor; (2) the electromagnetic force is proportional to the rate of change of the field; (3) the direction of the induced electromagnetic force depends on the orientation of the field. *electrolysis* (1) the amount of chemical change during electrolysis is proportional to the charge passing through the liquid; (2) the amount of chemical change produced in a substance by a given amount of electricity is proportional to the electrochemical equivalent of that substance.

Fermat's principle

principle that a ray of light, or other radiation, moves between two points along the path that takes the minimum time. The principle is named after French mathematician Pierre de Fermat, who used it to deduce the laws of reflection and refraction.

fermion

subatomic particle whose spin can only take values that are half-odd-integers, such as 1/2 or 3/2.

Fermions may be classified as leptons, such as the electron, and hadrons, such as the proton, neutron, mesons, and so on. All elementary particles are either fermions or bosons.

The exclusion principle, formulated by Austrian–born US physicist Wolfgang Pauli in 1925, asserts that no two fermions in the same system (such as an atom) can possess the same position, energy state, spin, or other quantized property.

ferromagnetism

form of magnetism that can be acquired in an external magnetic field and usually retained in its absence, so that ferromagnetic materials are used to make permanent magnets. A ferromagnetic material may therefore be said to have a high magnetic permeability and susceptibility (which depends upon temperature). Examples are iron, cobalt, nickel, and their alloys.

Ultimately, ferromagnetism is caused by spinning electrons in the atoms of the material, which act as tiny weak magnets. They align parallel to each other within small regions of the material to form domains, or areas of stronger magnetism. In an unmagnetized material, the domains are aligned at random so there is no overall magnetic effect. If a magnetic field is applied to that material, the domains align to point in the same direction, producing a strong overall magnetic effect. Permanent magnetism arises if the domains remain aligned after the external field is removed. Ferromagnetic materials exhibit hysteresis.

fibre optics

branch of physics dealing with the transmission of light and images through glass or plastic fibres known as optical fibres. Such fibres are now commonly used in both communications technology and medical investigations.

field

region of space in which an object exerts a force on another separate object because of certain properties they both possess. For example, there is a force of attraction between any two objects that have mass when one is in the gravitational field of the other.

Other fields of force include electric fields (caused by electric charges) and magnetic fields (caused by circulating electric currents), either of which can involve attractive or repulsive forces.

filter

circuit that transmits a signal of some frequencies better than others. A low-pass filter transmits signals of low frequency and also direct current; a high-pass filter transmits high-frequency signals; a band-pass filter transmits signals in a band of frequencies.

filter

device that absorbs some parts of the visible spectrum and transmits others. For example, a green filter will absorb or block all colours of the spectrum except green, which it allows to pass through. A yellow filter absorbs only light at the blue and violet end of the spectrum, transmitting red, orange, green, and yellow light.

fission

splitting of a heavy atomic nucleus into two or more major fragments. It is accompanied by the emission of two or three neutrons and the release of large amounts of nuclear energy.

Fission occurs spontaneously in nuclei of uranium-235, the main fuel used in nuclear reactors. However, the process can also be induced by bombarding nuclei with neutrons because a nucleus that has absorbed a neutron becomes unstable and soon splits. $^{235}_{92}U + ^{1}_{0}n \rightarrow ^{236}_{92}U \rightarrow 2$ nuclei + 2–3 neutrons + energy. The neutrons released spontaneously by the fission of uranium nuclei may therefore be used in turn to induce further fissions, setting up a chain reaction that must be controlled if it is not to result in a nuclear explosion. The minimum amount of fissile material that can undergo a continuous chain reaction is referred to as the critical mass.

Fleming's rules

memory aids used to recall the relative directions of the magnetic field, current, and motion in an electric generator or motor, using one's fingers. The three directions are represented by the thu*m*b (for *m*otion), *f*orefinger (for *f*ield), and se*c*ond finger (for conventional *c*urrent), all held at right angles to each other. The right hand is used for generators and the left for motors. The rules were devised by the English physicist John Fleming.

fluid mechanics

study of the behaviour of fluids (liquids and gases) at rest and in motion. Fluid mechanics is important in the study of the weather, the design of aircraft and road vehicles, and in industries, such as the chemical industry, which deal with flowing liquids or gases.

fluorescence

short-lived luminescence (a glow not caused by high temperature). Phosphorescence lasts a little longer.

Fluorescence is used in strip and other lighting, and was developed rapidly during World War II because it was a more efficient means of illumination than the incandescent lamp. Recently, small bulb-size fluorescence lamps have reached the market. It is claimed that, if widely used, their greater efficiency could reduce demand for electricity. Other important applications are in fluorescent screens for television and cathode-ray tubes.

FM

abbreviation for *frequency modulation.*

focal length or focal distance

distance from the centre of a lens or curved mirror to the focal point. For a concave mirror or convex lens, it is the distance at which rays of light parallel to the principal axis of the mirror or lens are brought to a focus (for a mirror, this is half the radius of curvature). For a convex mirror or concave lens, it is the distance from the centre to the point from which rays of light originally parallel to the principal axis of the mirror or lens diverge after being reflected or refracted.

With lenses, the greater the power (measured in dioptres) of the lens, the shorter its focal length. The human eye has a lens of adjustable focal length to allow the light from objects of varying distance to be focused on the retina.

focus or focal point
point at which light rays converge, or from which they appear to diverge. Other electromagnetic rays, such as microwaves, and sound waves may also be brought together at a focus. Rays parallel to the principal axis of a lens or mirror are converged at, or appear to diverge from, the principal focus.

force
any influence that tends to change the state of rest or the uniform motion in a straight line of a body. The action of an unbalanced or resultant force results in the acceleration of a body in the direction of action of the force, or it may, if the body is unable to move freely, result in its deformation (see Hooke's law). Force is a vector quantity, possessing both magnitude and direction; its SI unit is the newton.

speed and distance
In order to understand movement and what causes it, we need to be able to describe it. Speed is a measure of how fast something is moving. Speed is measured by dividing the distance travelled by the time taken to travel that distance. Hence speed is distance moved in unit time. Speed is a scalar quantity in which the direction of travel is not important, only the rate of travel. It is often useful to represent motion using a graph. Plotting distance against time in a distance–time graph enables one to calculate the total distance travelled. The gradient of the graph represents the speed at a particular point, the instantaneous speed. A straight line on the distance–time graph corresponds to a constant speed. A form of this graph that shows the stages on the journey is called a travel graph. A speed–time graph plots speed against time. It shows the instantaneous speed at each point. When the graph is horizontal, the object is stationary because the speed is zero.

velocity and acceleration
Velocity is the speed of an object in a given direction. Velocity is therefore a vector quantity, in which both magnitude and direction of movement must be taken into account. Acceleration is the rate of change of velocity with time. This is also a vector quantity. Acceleration happens when there is a change in speed, or a change in direction, or a change in speed and direction.

forces and motion
Galileo discovered that a body moving on a perfectly smooth horizontal surface would neither speed up nor slow down. All moving bodies continue moving

with the same velocity unless a force is applied to cause an acceleration. The reason we appear to have to push something to keep it moving with constant velocity is because of frictional forces acting on all moving objects on Earth. Friction occurs when two solid surfaces rub on each other; for example, a car tyre in contact with the ground. Friction opposes the relative motion of the two objects in contact and acts to slow the velocity of the moving object. A force is required to push the moving object and to cancel out the frictional force. If the forces combine to give a net force of zero, the object will not accelerate but will continue moving at constant velocity. A resultant force is a single force acting on a particle or body whose effect is equivalent to the combined effects of two or more separate forces. Galileo's work was developed by Isaac Newton. According to Newton's second law of motion, the magnitude of a resultant force is equal to the rate of change of momentum of the body on which it acts; the force F producing an acceleration a m s^{-2} on a body of mass m kilograms is therefore given by:

$$F = ma$$

Thus Newton's second law states that change of momentum is proportional to the size of the external force and takes place in the direction in which the force acts. Momentum is a function both of the mass of a body and of its velocity. This agrees with our experience, because the idea of force is derived from muscular effort, and we know that we have to exert more strength to stop the motion of a heavy body than a light one, just as we have to exert more strength to stop a rapidly moving body than a slowly moving one. Force, then, is measured by change of momentum, momentum being equal to mass multiplied by velocity. (See also Newton's laws of motion.) Newton's third law of motion states that if a body A exerts a force on a body B, then body B exerts an equal force on body A but in the opposite direction. This equal and opposite force is called a reaction force.

free fall and terminal velocity
Galileo also established that freely falling bodies, heavy or light, have the same, constant acceleration and that this acceleration is due to gravity. This acceleration, due to the gravitational force exerted by the Earth, is also known as the acceleration of free fall. It has a value of 10 m s^{-2}. However, air resistance acts when a body falls through the air. This increases greatly as an object's velocity increases, so the object tends to reach a terminal velocity. It then continues to fall with this same velocity (it has stopped accelerating because its weight is cancelled out by air resistance) until it reaches the ground. The acceleration due to gravity can be measured using a pendulum.

statics
Statics is the branch of mechanics concerned with the behaviour of bodies at rest or moving with constant velocity. The forces acting on the body under these circumstances cancel each other out; that is, the forces are in equilibrium.

force ratio
magnification of a force by a machine; see mechanical advantage.

forces, fundamental
four fundamental interactions currently known to be at work in the physical universe. There are two long-range forces: the *gravitational force*, or *gravity*, which keeps the planets in orbit around the Sun, and acts between all particles that have mass; and the *electromagnetic force*, which stops solids from falling apart, and acts between all particles with electric charge. There are two very short-range forces which operate over distances comparable with the size of the atomic nucleus: the *weak nuclear force*, responsible for the reactions that fuel the Sun and for the emission of beta particles by some particles; and the *strong nuclear force*, which binds together the protons and neutrons in the nuclei of atoms. The relative strengths of the four forces are: strong, 1; electromagnetic, 10^{-2}; weak, 10^{-6}; gravitational, 10^{-40}.

By 1971, the US physicists Steven Weinberg and Sheldon Glashow, the Pakistani physicist Abdus Salam, and others had developed a theory that suggested that the weak and electromagnetic forces were aspects of a single force called the *electroweak force*; experimental support came from observation at CERN in the 1980s. Physicists are now working on theories to unify all four forces. See supersymmetry.

frequency
number of periodic oscillations, vibrations, or waves occurring per unit of time. The SI unit of frequency is the hertz (Hz), one hertz being equivalent to one cycle per second. Frequency is related to wavelength and velocity by the relationship:

$$f = \frac{v}{\lambda}$$

where f is frequency, v is velocity, and λ is wavelength. Frequency is the reciprocal of the period T:

$$f = \frac{1}{T}$$

Human beings can hear sounds from objects vibrating in the range 20–15,000 Hz. Ultrasonic frequencies well above 15,000 Hz can be detected by such mammals as bats. Infrasound (low-frequency sound) can be detected by some animals and birds. Pigeons can detect sounds as low as 0.1 Hz; elephants communicate using sounds as low as 1 Hz. *Frequency modulation* (FM) is a method of transmitting radio signals in which the frequency of the *carrier wave* is changed and then decoded.

One kilohertz (kHz) equals 1,000 hertz; one megahertz (MHz) equals 1,000,000 hertz.

frequency modulation (FM)

method by which radio waves are altered for the transmission of broadcasting signals. FM varies the frequency of the carrier wave in accordance with the signal being transmitted. Its advantage over AM (amplitude modulation) is its better signal-to-noise ratio. It was invented by the US engineer Edwin Armstrong.

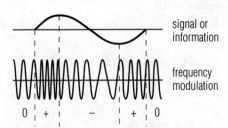

In FM radio transmission, the frequency of the carrier wave is modulated, rather than its amplitude (as in AM broadcasts). The FM system is not affected by the many types of interference that change the amplitude of the carrier wave, and so provides better quality reception than AM broadcasts.

friction

force that opposes the relative motion of two bodies in contact. The *coefficient of friction* is the ratio of the force required to achieve this relative motion to the force pressing the two bodies together.

Friction is greatly reduced by the use of lubricants such as oil, grease, and graphite. Air bearings are now used to minimize friction in high-speed rotational machinery. In other instances friction is deliberately increased by making the surfaces rough – for example, brake linings, driving belts, soles of shoes, and tyres.

fuel cell

cell converting chemical energy directly to electrical energy. It works on the same principle as a battery but is continually fed with fuel, usually hydrogen. Fuel cells are silent and reliable (no moving parts) but expensive to produce.

Hydrogen is passed over an electrode (usually nickel or platinum) containing a catalyst, which strips electrons off the atoms. These pass through an external circuit while hydrogen ions (charged atoms) pass through an electrolyte to another electrode, over which oxygen is passed. Water is formed at this electrode (as a by-product) in a chemical reaction involving electrons, hydrogen ions, and oxygen atoms. If the spare heat also produced is used for hot water and space heating, 80% efficiency in fuel is achieved.

fundamental constant

physical quantity that is constant in all circumstances throughout the whole universe. Examples are the electric charge of an electron, the speed of light, Planck's constant, and the gravitational constant.

fundamental forces

see forces, fundamental.

fundamental particle

another term for *elementary particle*.

fusion

fusing of the nuclei of light elements, such as hydrogen, into those of a heavier element, such as helium. The resultant loss in their combined mass is converted into energy. Stars and thermonuclear weapons are powered by nuclear fusion.

Very high temperatures and pressures are thought to be required in order for fusion to take place. Under these conditions the atomic nuclei can approach each other at high speeds and overcome the mutual repulsion of their positive charges. At very close range another force, the strong nuclear force, comes into play, fusing the particles together to form a larger nucleus. As fusion is accompanied by the release of large amounts of energy, the process might one day be harnessed to form the basis of commercial energy production. So far no successful fusion reactor – one able to produce the required conditions and contain the reaction – has been built. However, an important step along the road to fusion power was taken in 1991. In an experiment that lasted 2 seconds, a 1.7 megawatt pulse of power was produced by the Joint European Torus (JET) at Culham, Oxfordshire, UK. This was the first time that a substantial amount of fusion power had been produced in a controlled experiment, as opposed to a bomb. In 1997 JET produced a record 21 megajoule of fusion power, and tested the first large-scale plant of the type needed to supply and process tritium in a future fusion power station.

Pinch fusion occurs when an electric current is passed through a plasma and generates a magnetic field which constricts the particles.

gain

ratio of the amplitude of the output signal produced by an amplifier to that of the input signal. In a voltage amplifier the voltage gain is the ratio of the output voltage to the input voltage; in an inverting operational amplifier (op-amp) it is equal to the ratio of the resistance of the feedback resistor to that of the input resistor.

gamma radiation

very high-frequency electromagnetic radiation, similar in nature to X-rays but of shorter wavelength, emitted by the nuclei of radioactive substances during decay or by the interactions of high-energy electrons with matter. Cosmic gamma rays have been identified as coming from pulsars, radio galaxies, and quasars, although they cannot penetrate the Earth's atmosphere.

Gamma rays are stopped only by direct collision with an atom and are therefore very penetrating; they can, however, be stopped by about 4 cm/1.5 in of lead or by a very thick concrete shield. They are less ionizing in their effect than alpha and beta particles, but are dangerous nevertheless because they can penetrate deeply into body tissues such as bone marrow. They are not deflected by either magnetic or electric fields. Gamma radiation is used to kill bacteria and other micro-organisms, sterilize medical devices, and change the molecular structure of plastics to modify their properties (for example, to improve their resistance to heat and abrasion).

gas

form of matter, such as air, in which the molecules move randomly in otherwise empty space, filling any size or shape of container into which the gas is put.

A sugar-lump sized cube of air at room temperature contains 30 trillion molecules moving at an average speed of 500 m per second (1,800 kph/ 1,200 mph). Gases can be liquefied by cooling, which lowers the speed of the molecules and enables attractive forces between them to bind them together.

gas constant

constant R that appears in the equation $PV = nRT$, which describes how the pressure P, volume V, and temperature T of an ideal gas are related (n is the amount of gas in moles). This equation combines Boyle's law and Charles's law.

R has a value of 8.3145 joules per kelvin per mole.

gas laws

physical laws concerning the behaviour of gases. They include Boyle's law and Charles's law, which are concerned with the relationships between the pressure, temperature, and volume of an ideal (hypothetical) gas. These two laws can be combined to give the *general* or *universal gas law*, which may be expressed as:

$$\frac{(\text{pressure} \times \text{volume})}{\text{temperature} = \text{constant}} = \text{constant}$$

Van der Waals' law includes corrections for the nonideal behaviour of real gases.

gauge boson or field particle

any of the particles that carry the four fundamental forces of nature (see forces, fundamental). Gauge bosons are elementary particles that cannot be subdivided, and include the photon, the graviton, the gluons, and the weakons.

gear

toothed wheel that transmits the turning movement of one shaft to another shaft. Gear wheels may be used in pairs, or in threes if both shafts are to turn in the same direction. The gear ratio – the ratio of the number of teeth on the two wheels – determines the torque ratio, the turning force on the output shaft compared with the turning force on the input shaft. The ratio of the angular velocities of the shafts is the inverse of the gear ratio.

The common type of gear for parallel shafts is the *spur gear*, with straight teeth parallel to the shaft axis. The *helical gear* has teeth cut along sections of a helix or corkscrew shape; the double form of the helix gear is the most efficient for energy transfer. *Bevel gears*, with tapering teeth set on the base of a cone, are used to connect intersecting shafts. *See illustration on page 196.*

Geiger counter

any of a number of devices used for detecting nuclear radiation and/or measuring its intensity by counting the number of ionizing particles produced (see radioactivity). It detects the momentary current that passes between electrodes in a suitable gas when a nuclear particle or a radiation pulse causes the ionization of that gas. The electrodes are connected to electronic devices that enable the number of particles passing to be measured. The increased frequency of measured particles indicates the intensity of radiation. The device is named after the German physicist Hans Geiger.

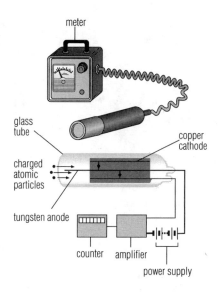

A Geiger–Müller counter detects and measures ionizing radiation (alpha particles, beta particles, and gamma rays) emitted by radioactive materials. Any incoming radiation creates ions (charged particles) within the counter, which are attracted to the anode and cathode to create a measurable electric current.

generator

machine that produces electrical energy from mechanical energy, as opposed to an electric motor, which does the opposite. A simple generator (known as a dynamo in the UK) consists of a wire-wound coil (armature) that is rotated between the poles of a permanent magnet. The movement of the wire in the magnetic field induces a current in the coil by electromagnetic induction, which can be fed by means of a commutator as a continuous direct current into an external circuit. Slip rings instead of a commutator produce an alternating current, when the generator is called an alternator.

gluon

gauge boson that carries the strong nuclear force, responsible for binding quarks together to form the strongly interacting subatomic particles known as hadrons. There are eight kinds of gluon.

Gluons cannot exist in isolation; they are believed to exist in balls ('glueballs') that behave as single particles. Glueballs may have been detected at CERN in 1995 but further research is required to confirm their existence.

grand unified theory

sought-for theory that would combine the theory of the strong nuclear force (called quantum chromodynamics) with the theory of the weak nuclear and electromagnetic forces. The search for the grand unified theory is part of a larger

ratio 1:2	ratio 2:1	ratio 2:3
(one turn of the driver to every two of the driven)	(two turns of the driver to every one of the driven)	(two turns of the driver to every three of the driven)

Gear ratio is calculated by dividing the number of teeth on the driver gear by the number of teeth on the driven gear (gear ratio = $\frac{driver}{driven}$); the idler gears are ignored. Idler gears change the direction of rotation but do not affect speed. A high driven to driver ratio (middle) is a speed-reducing ratio.

programme seeking a unified field theory, which would combine all the forces of nature (including gravity) within one framework.

gravitational field
region around a body in which other bodies experience a force due to its gravitational attraction. The gravitational field of a massive object such as the Earth is very strong and easily recognized as the force of gravity, whereas that of an object of much smaller mass is very weak and difficult to detect. Gravitational fields produce only attractive forces.

gravitational force or gravity
one of the four fundamental forces of nature, the other three being the electromagnetic force, the weak nuclear force, and the strong nuclear force. The gravitational force is the weakest of the four forces, but it acts over great distances. The particle that is postulated as the carrier of the gravitational force is the graviton.

gravitational lensing
bending of light by a gravitational field, predicted by German-born US physicist Albert Einstein's general theory of relativity. The effect was first detected in

1917 when the light from stars was found to be bent as it passed the totally eclipsed Sun. More remarkable is the splitting of light from distant quasars into two or more images by intervening galaxies. In 1979 the first double image of a quasar produced by gravitational lensing was discovered and a quadruple image of another quasar was later found.

graviton
gauge boson that is the postulated carrier of the gravitational force.

gravity
force of attraction that arises between objects by virtue of their masses. On Earth, gravity is the force of attraction between any object in the Earth's gravitational field and the Earth itself.

One of the earliest gravitational experiments was undertaken by Nevil Maskelyne in 1774 and involved the measurement of the attraction of Mount Schiehallion (Scotland) on a plumb bob.

Henry Cavendish, in 1797, obtained the gravitational constant G with a value of 6.6×10^{-11} N m^2 kg^{-2}. The value generally used today is 6.6720×10^{-11} N m^2 kg^{-2}.

Newton's laws
According to Isaac Newton's law of gravitation, all objects fall to Earth with the same acceleration, regardless of mass. For an object of mass m_1 at a distance r from the centre of the Earth (mass m_2), the gravitational force of attraction F equals $Gm_1 m_2 / r^2$, where G is the gravitational constant. However, according to Newton's second law of motion, F also equals $m_1 g$, where g is the acceleration due to gravity; therefore $g = Gm_2 / r^2$ and is independent of the mass of the object; at the Earth's surface it equals 9.806 m per second per second.

relativity
Albert Einstein's general theory of relativity treats gravitation not as a force but as the curvature of space-time around a body. Relativity predicts the bending of light and the red shift of light in a gravitational field; both have been observed. Another prediction of relativity is ***gravitational waves***, which should be produced when massive bodies are violently disturbed. These waves are so weak that they have not yet been detected with certainty, although observations of a pulsar (which emits energy at regular intervals) in orbit around another star have shown that the stars are spiralling together at the rate that would be expected if they were losing energy in the form of gravitational waves.

hadron
subatomic particle that experiences the strong nuclear force. Each is made up of two or three indivisible particles called quarks. The hadrons are grouped into the baryons (protons, neutrons, and hyperons), consisting of three quarks, and the mesons, consisting of two quarks.

half-life

during radioactive decay, the time in which the activity of a radioactive source decays to half its original value. In theory, the decay process is never complete and there is always some residual radioactivity. For this reason, the half-life of a radioactive isotope is measured, rather than the total decay time. It may vary from millionths of a second to billions of years.

Radioactive substances decay exponentially; thus the time taken for the first 50% of the isotope to decay will be the same as the time taken by the next 25%, and by the 12.5% after that, and so on. For example, carbon-14 takes about 5,730 years for half the material to decay; another 5,730 for half of the remaining half to decay; then 5,730 years for half of that remaining half to decay, and so on. Plutonium-239, one of the most toxic of all radioactive substances, has a half-life of about 24,000 years.

heat

form of energy possessed by a substance by virtue of the vibrational movement (kinetic energy) of its molecules or atoms. Heat energy is transferred by conduction, convection, and radiation. It always flows from a region of higher temperature (heat intensity) to one of lower temperature. Its effect on a substance may be simply to raise its temperature, or to cause it to expand, melt (if a solid), vaporize (if a liquid), or increase its pressure (if a confined gas).

measurement

Quantities of heat are usually measured in units of energy, such as joules (J) or calories (cal). The specific heat of a substance is the ratio of the quantity of heat required to raise the temperature of a given mass of the substance through a given range of temperature to the heat required to raise the temperature of an equal mass of water through the same range. It is measured by a calorimeter.

conduction, convection, and radiation

Conduction is the passing of heat along a medium to neighbouring parts with no visible motion accompanying the transfer of heat; for example, when the whole length of a metal rod is heated when one end is held in a fire. Convection is the transmission of heat through a fluid (liquid or gas) in currents; for example, when the air in a room is warmed by a fire or radiator. Radiation is heat transfer by infrared rays. It can pass through a vacuum, travels at the same speed as light, and can be reflected and refracted; for example, heat reaches the Earth from the Sun by radiation. For the transformation of heat, see thermodynamics.

heat capacity

quantity of heat required to raise the temperature of an object by one degree. The *specific heat capacity* of a substance is the heat capacity per unit of mass, measured in joules per kilogram per kelvin ($J kg^{-1} K^{-1}$).

Higgs boson or Higgs particle

postulated elementary particle whose existence would explain why particles have mass. The current theory of elementary particles, called the standard model, cannot explain how mass arises. To overcome this difficulty, Peter Higgs of the University of Edinburgh and Thomas Kibble of Imperial College, London proposed in 1964 a new particle that binds to other particles and gives them their mass. Physicists in Geneva announced in September 2000 that they may have found evidence that they successfully created a Higgs boson. This would, if confirmed, be the first successful creation of the particle.

Hooke's law

law stating that the deformation of a body is proportional to the magnitude of the deforming force, provided that the body's elastic limit (see elasticity) is not exceeded. If the elastic limit is not reached, the body will return to its original size once the force is removed. The law was discovered by English physicist Robert Hooke in 1676.

For example, if a spring is stretched by 2 cm by a weight of 1 N, it will be stretched by 4 cm by a weight of 2 N, and so on; however, once the load exceeds the elastic limit for the spring, Hooke's law will no longer be obeyed and each successive increase in weight will result in a greater extension until finally the spring breaks.

humidity

quantity of water vapour in a given volume of the atmosphere (absolute humidity), or the ratio of the amount of water vapour in the atmosphere to the saturation value at the same temperature (relative humidity). At dew point the relative humidity is 100% and the air is said to be saturated. Condensation (the conversion of vapour to liquid) may then occur. Relative humidity is measured by various types of hygrometer.

hydrodynamics

branch of physics dealing with fluids (liquids and gases) in motion.

hydrometer

instrument used to measure the relative density of liquids (the density compared with that of water). A hydrometer consists of a thin glass tube ending in a sphere that leads into a smaller sphere, the latter being weighted so that the hydrometer floats upright, sinking deeper into less dense liquids than into denser liquids. Hydrometers are used in brewing and to test the strength of acid in car batteries.

The hydrometer is based on Archimedes' principle.

hydrostatics

branch of statics dealing with fluids in equilibrium – that is, in a static condition. Practical applications include shipbuilding and dam design.

hypercharge

property of certain elementary particles, analogous to electric charge, that accounts for the absence of some expected behaviour (such as certain decays).

Protons and neutrons, for example, have a hypercharge of +1, whereas a π meson has a hypercharge of 0.

image

picture or appearance of a real object, formed by light that passes through a lens or is reflected from a mirror. If rays of light actually pass through an image, it is called a *real image*. Real images, such as those produced by a camera or projector lens, can be projected onto a screen. An image that cannot be projected onto a screen, such as that seen in a flat mirror, is known as a *virtual image*.

impedance

total opposition of a circuit to the passage of alternating electric current. It has the symbol Z. It includes the resistance R and the reactance X (caused by capacitance or inductance); the impedance can then be found using the equation $Z^2 = R^2 + X^2$.

incandescence

emission of light from a substance in consequence of its high temperature. The colour of the emitted light from liquids or solids depends on their temperature, and for solids generally the higher the temperature the whiter the light. Gases may become incandescent through ionizing radiation, as in the glowing vacuum discharge tube.

The oxides of cerium and thorium are highly incandescent and for this reason are used in gas mantles. The light from an electric filament lamp is due to the incandescence of the filament, rendered white-hot when a current passes through it.

inductance

phenomenon in which a changing current in a circuit builds up a magnetic field which induces an electromotive force either in the same circuit and opposing the current (self-inductance) or in another circuit (mutual inductance). The SI unit of inductance is the henry (symbol H).

A component designed to introduce inductance into a circuit is called an inductor (sometimes inductance) and is usually in the form of a coil of wire. The energy stored in the magnetic field of the coil is proportional to its inductance and the current flowing through it. See electromagnetic induction.

inductor

device included in an electrical circuit because of its inductance.

inertia

tendency of an object to remain in a state of rest or uniform motion until an external force is applied, as described by Isaac Newton's first law of motion (see Newton's laws of motion).

infrared radiation

invisible electromagnetic radiation of wavelength between about 0.75 micrometres and 1 millimetre – that is, between the limit of the red end of the visible spectrum and the shortest microwaves. All bodies above the absolute zero of temperature absorb and radiate infrared radiation. Infrared radiation is used in medical photography and treatment, and in industry, astronomy, and criminology.

Infrared absorption spectra are used in chemical analysis, particularly for organic compounds. Objects that radiate infrared radiation can be photographed or made visible in the dark on specially sensitized emulsions. This is important for military purposes and in detecting people buried under rubble. The strong absorption by many substances of infrared radiation is a useful method of applying heat.

insulator

any poor conductor of heat, sound, or electricity. Most substances lacking free (mobile) electrons, such as non-metals, are electrical or thermal insulators. Usually, devices of glass or porcelain, called insulators, are used for insulating and supporting overhead wires.

integrated circuit (IC) or silicon chip

miniaturized electronic circuit produced on a single crystal, or chip, of a semiconducting material – usually silicon. It may contain many millions of components and yet measure only 5 mm/0.2 in square and 1 mm/0.04 in thick. The IC is encapsulated within a plastic or ceramic case, and linked via gold wires to metal pins with which it is connected to a printed circuit board and the other components that make up such electronic devices as computers and calculators.

intensity

power (or energy per second) per unit area carried by a form of radiation or wave motion. It is an indication of the concentration of energy present and, if measured at varying distances from the source, of the effect of distance on this.

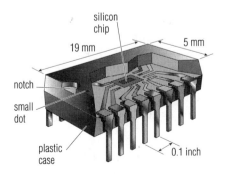

An integrated circuit (IC), or silicon chip.

For example, the intensity of light is a measure of its brightness, and may be shown to diminish with distance from its source in accordance with the inverse square law (its intensity is inversely proportional to the square of the distance).

interference
phenomenon of two or more wave motions interacting and combining to produce a resultant wave of larger or smaller amplitude (depending on whether the combining waves are in or out of phase with each other).

Interference of white light (multiwavelength) results in spectral coloured fringes; for example, the iridescent colours of oil films seen on water or soap bubbles (demonstrated by Newton's rings). Interference of sound waves of similar frequency produces the phenomenon of beats, often used by musicians when tuning an instrument. With monochromatic light (of a single wavelength), interference produces patterns of light and dark bands. This is the basis of holography, for example. Interferometry can also be applied to radio waves, and is a powerful tool in modern astronomy.

intermediate vector boson
member of a group of elementary particles, W^+, W^-, and Z, which mediate the weak nuclear force. This force is responsible for, amongst other things, beta decay.

intermolecular force or van der Waals' force
force of attraction between molecules. Intermolecular forces are relatively weak; hence simple molecular compounds are gases, liquids, or low-melting-point solids.

inverse square law
statement that the magnitude of an effect (usually a force) at a point is inversely proportional to the square of the distance between that point and the object exerting the force.

Light, sound, electrostatic force (Coulomb's law), and gravitational force (Newton's law) all obey the inverse square law.

ion
atom, or group of atoms, that is either positively charged (cation) or negatively charged (anion), as a result of the loss or gain of electrons during chemical reactions or exposure to certain forms of radiation. In solution or in the molten state, ionic compounds such as salts, acids, alkalis, and metal oxides conduct electricity. These compounds are known as electrolytes.

Ions are produced during electrolysis.

ionizing radiation
radiation that knocks electrons from atoms during its passage, thereby leaving ions in its path. Alpha and beta particles are far more ionizing in their effect than are neutrons or gamma radiation.

ion plating

method of applying corrosion-resistant metal coatings. The article is placed in argon gas, together with some coating metal, which vaporizes on heating and becomes ionized (acquires charged atoms) as it diffuses through the gas to form the coating. It has important applications in the aerospace industry.

isotope

one of two or more atoms that have the same atomic number (same number of protons), but which contain a different number of neutrons, thus differing in their atomic mass (see relative atomic mass). They may be stable or radioactive (see radioisotope), naturally occurring, or synthesized. For example, hydrogen has the isotopes ^{2}H (deuterium) and ^{3}H (tritium). The term was coined by English chemist Frederick Soddy, pioneer researcher in atomic disintegration.

Elements at the lower end of the periodic table have atoms with roughly the same number of protons as neutrons. These elements are called *stable isotopes*. The stable isotopes of oxygen include ^{16}O, ^{17}O, and ^{18}O. Elements with high atomic mass numbers have many more neutrons than protons and are therefore less stable. It is these isotopes that are more prone to radioactive decay. One example is ^{238}U, uranium-238.

J

symbol for *joule*, the SI unit of energy.

joule

SI unit (symbol J) of work and energy, replacing the calorie (one calorie equals 4.2 joules).

It is defined as the work done (energy transferred) by a force of one newton acting over one metre. It can also be expressed as the work done in one second by a current of one ampere at a potential difference of one volt. One watt is equal to one joule per second.

Kelvin scale

temperature scale used by scientists. It begins at absolute zero (−273.15°C) and increases in kelvin, the same degree intervals as the Celsius scale; that is, 0°C is the same as 273.15 K and 100°C is 373.15 K. It is named after the Scottish physicist William Thomson, 1st Baron Kelvin.

Kennelly–Heaviside layer

former term for the E layer of the ionosphere.

kilowatt-hour

commercial unit of electrical energy (symbol kWh), defined as the work done by a power of 1,000 watts in one hour and equal to 3.6 megajoules. It is used to calculate the cost of electrical energy taken from the domestic supply.

kinetic energy

energy of a body resulting from motion. It is contrasted with potential energy. The kinetic energy of a moving body is equal to the work that would have to be done in bringing that body to rest, and is dependent upon both the body's mass and speed. The kinetic energy in joules of a mass m kg travelling with speed v m s^{-1} is given by the formula:

$$KE = \frac{mv^2}{2}$$

All atoms and molecules possess some amount of kinetic energy because they are all in some state of motion (see kinetic theory). Adding heat energy to a substance increases the mean kinetic energy and hence the mean speed of its constituent molecules – a change that is reflected as a rise in the temperature of that substance.

kinetics

branch of dynamics dealing with the action of forces producing or changing the motion of a body; *kinematics* deals with motion without reference to force or mass.

kinetic theory

theory describing the physical properties of matter in terms of the behaviour – principally movement – of its component atoms or molecules. The temperature of a substance is dependent on the velocity of movement of its constituent particles, increased temperature being accompanied by increased movement. A gas consists of rapidly moving atoms or molecules and, according to kinetic theory, it is their continual impact on the walls of the containing vessel that accounts for the pressure of the gas. The slowing of molecular motion as temperature falls, according to kinetic theory, accounts for the physical properties of liquids and solids, culminating in the concept of no molecular motion at absolute zero (0 K/–273.15°C).

By making various assumptions about the nature of gas molecules, it is possible to derive from the kinetic theory the gas laws (such as Avogadro's hypothesis, Boyle's law, and Charles's law).

Large Electron Positron Collider LEP

the world's largest particle accelerator, in operation 1989–2000 at the CERN laboratories near Geneva in Switzerland. It occupies a tunnel 3.8 m/12.5 ft wide and 27 km/16.7 mi long, which is buried 180 m/590 ft underground and forms a ring consisting of eight curved and eight straight sections. In June 1996, LEP resumed operation after a £210 million upgrade. The upgraded machine, known as LEP2, generated collision energy of 161 gigaelectron volts.

Electrons and positrons enter the ring after passing through the Super Proton Synchrotron accelerator. They travel in opposite directions around the ring, guided by 3,328 bending magnets and kept within tight beams by 1,272 focusing magnets. As they pass through the straight sections, the particles are accelerated by a pulse of radio energy. Once sufficient energy is accumulated, the beams are allowed to collide. Four giant detectors are used to study the resulting shower of particles.

laser acronym for light amplification by stimulated emission of radiation

device for producing a narrow beam of light, capable of travelling over vast distances without dispersion, and of being focused to give enormous power densities (10^8 watts per cm^2 for high-energy lasers). The laser operates on a principle similar to that of the maser (a high-frequency microwave amplifier or oscillator). The uses of lasers include communications (a laser beam can carry much more information than can radio waves), cutting, drilling, welding, satellite tracking, medical and biological research, and surgery. Sound wave vibrations from the window glass of a room can be picked up by a reflected laser beam. Lasers are also used as entertainment in theatres, concerts, and light shows.

laser material

Any substance in which the majority of atoms or molecules can be put into an excited energy state can be used as laser material. Many solid, liquid, and gaseous substances have been used, including synthetic ruby crystal (used for the first extraction of laser light in 1960, and giving a high-power pulsed output) and a helium–neon gas mixture, capable of continuous operation, but at a lower power.

applications

Carbon dioxide gas lasers (CO_2 lasers) can produce a beam of 100 watts or more power in the infrared (wavelength 10.6 μm) and this has led to an important commercial application, the cutting of material for suits and dresses in hundreds of thicknesses at a time. Dye lasers, in which complex organic dyes in solution are the lasing material, can be tuned to produce light of any chosen wavelength over a range of a sizeable fraction of the visible spectrum.

photon emission

An atom can emit a photon of light (an elementary wave train) if it has somehow gained enough energy to do so; if it has gained this energy it is said to be in an excited state and this can occur, for example, by collision with another atom or by irradiation with light of suitable wavelength. The process of providing the atoms with energy is called 'pumping'. Normally the atom will emit its photon very quickly (in less than 10^{-6} s) and at random (spontaneous emission), but if a photon of the same wavelength passes while the atom is still in an excited state the atom will emit its photon in phase with the passing photon (stimulated emission). In a laser it is arranged that this process takes place in a manner that causes a rapid build-up of light intensity.

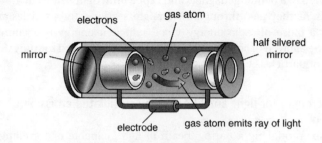

electrons   gas atom

mirror

half silvered mirror

electrode   gas atom emits ray of light

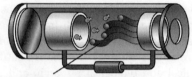

light ray hits another energized
atom causing more light to be emitted

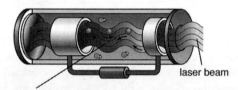

laser beam

light rays bounce between the
mirrors causing a build up of light

In a gas laser, electrons moving between the electrodes pass energy to gas atoms. An energized atom emits a ray of light. The ray hits another energized atom causing it to emit a further light ray. The rays bounce between mirrors at each end causing a build-up of light. Eventually it becomes strong enough to pass through the half-silvered mirror at one end, producing a laser beam.

helium–neon lasers

The helium–neon laser is the commonest and cheapest kind; it consists of a sealed glass tube containing a mixture of helium and neon gases at low pressure and with mirrors sealed onto the tube at either end. An electrical discharge is passed through the tube from two sealed-in electrodes. The energy of the discharge 'pumps' the neon atoms, and photons of wavelength 0.6328×10^{-6} m are emitted (red light). The light bounces between the mirrors and is amplified at each pass. One of the mirrors is slightly transparent to allow the pencil beam of red light to emerge.

laser coherence

Laser light is very coherent and it can be used to demonstrate a variety of interference effects conveniently; it is used for holography and for accurate length measurement by interferometry. On account of its coherence, laser light can be accurately collimated; if a collimated beam 1 m/3.3 ft in diameter is made it will spread by diffraction at an angle much less than 1′ of arc. This has made it possible to send laser pulses to the Moon; by detecting the pulses reflected back from mirrors placed on the Moon very accurate measurements of the Moon's distance can be made. Quite different applications are for concentrated heat in surgery and welding, and for many purposes in pure science. Lasers can be made to produce very short pulses of light (10^{-11} s) with very intense peak power (10^{14} watts); such pulses are a possible means for initiating the fusion of light elements to produce energy, in the same way that the energy of the Sun is produced.

atom laser
US scientists unveiled the first atom laser in 1997. It emits a new type of matter with atoms acting like lightwaves, and the inventors predicted their new laser will lead to advances in computer chips and navigational equipment.

latent heat
heat absorbed or released by a substance as it changes state (for example, from solid to liquid) at constant temperature and pressure.

lens
piece of a transparent material, such as glass, with two polished surfaces – one concave or convex, and the other plane, concave, or convex – that modifies rays of light. A convex lens brings rays of light together; a concave lens makes the rays diverge. Lenses are essential to spectacles, microscopes, telescopes, cameras, and almost all optical instruments.

The image formed by a single lens suffers from several defects or aberrations, notably *spherical aberration* in which an image becomes blurred, and *chromatic aberration* in which an image in white light tends to have coloured edges. Aberrations are corrected by the use of compound lenses, which are built up from two or more lenses of different refractive index.

Lenz's law
law stating that the direction of an electromagnetically induced current (generated by moving a magnet near a wire or a wire in a magnetic field) will be such as to oppose the motion producing it. It is named after the German physicist Heinrich Friedrich Lenz (1804–1865), who announced it in 1833.

lepton
any of a class of elementary particles that are not affected by the strong nuclear force. The leptons comprise the electron, muon, and tau, and their neutrinos (the electron neutrino, muon neutrino, and tau neutrino), as well as their six antiparticles.

In July 2000, researchers at Fermilab, in the USA, amassed the first direct evidence for the existence of the tau lepton.

lever

simple machine consisting of a rigid rod pivoted at a fixed point called the fulcrum, used for shifting or raising a heavy load or applying force. Levers are classified into orders according to where the effort is applied, and the load-moving force developed, in relation to the position of the fulcrum.

A *first-order* lever has the load and the effort on opposite sides of the fulcrum – for example, a see-saw or pair of scissors. A *second-order* lever has the load and the effort on the same side of the fulcrum, with the load nearer the fulcrum – for example, nutcrackers or a wheelbarrow. A *third-order* lever has the effort nearer the fulcrum than the load, with both on the same side of it – for example, a pair of tweezers or tongs. The mechanical advantage of a lever is the ratio of load to effort, equal to the perpendicular distance of the effort's line of action from the fulcrum divided by the distance to the load's line of action. Thus tweezers, for instance, have a mechanical advantage of less than one.

light

electromagnetic waves in the visible range, having a wavelength from about 400 nanometres in the extreme violet to about 770 nanometres in the extreme red. Light is considered to exhibit particle and wave properties, and the fundamental particle, or quantum, of light is called the photon. The speed of light (and of all electromagnetic radiation) in a vacuum is approximately 300,000 km/186,000 mi per second, and is a universal constant denoted by c.

Isaac Newton was the first to discover, in 1666, that sunlight is composed of a mixture of light of different colours in certain proportions and that it could be separated into its components by dispersion. Before his time it was supposed that dispersion of light produced colour instead of separating already existing colours. The ancients believed that light travelled at infinite speed; its finite speed was first discovered by Danish astronomer Ole Römer in 1676.

modern theories

The emission of light from self-luminous bodies is an atomic phenomenon that was given a satisfactory explanation by Max Planck, Niels Bohr, and others in terms of the quantum theory. Similarly the absorption of light, and in particular the emission of electrons from metallic surfaces illuminated by light (photoelectric effect), was explained by the quantum theory. Certain properties of light, however, are explained only on the hypothesis that light is propagated as electromagnetic waves. Thus the quantum theory accounts for the photoelectric effect, while the electromagnetic wave theory accounts for the interference of light. The relation between these two theories can be approached in terms of Werner Heisenberg's uncertainty principle.

lightning

high-voltage electrical discharge between two charged rainclouds or between a cloud and the Earth, caused by the build-up of electrical charges. Air in the path of lightning ionizes (becomes conducting), and expands; the accompanying noise is heard as thunder. Currents of 20,000 amperes and temperatures of 30,000°C/54,000°F are common. Lightning causes nitrogen oxides to form in the atmosphere and approximately 25% of the atmospheric nitrogen oxides are formed in this way.

According to a 1997 US survey on lightning strength and frequency, using information gathered from satellite images and data from the US Lightning Detection Network, there are 70–100 lightning flashes per second worldwide, with an average peak current of 36 kiloamps.

lightning conductor

device that protects a tall building from lightning strike by providing an easier path for current to flow to earth than through the building. It consists of a thick copper strip of very low resistance connected to the ground below. A good connection to the ground is essential and is made by burying a large metal plate deep in the damp earth. In the event of a direct lightning strike, the current in the conductor may be so great as to melt or even vaporize the metal, but the damage to the building will nevertheless be limited.

linear accelerator or linac

type of particle accelerator in which the particles move along a straight tube. Particles pass through a linear accelerator only once – unlike those in a cyclotron or synchrotron (ring-shaped accelerators), which make many revolutions, gaining energy each time.

The world's longest linac is the Stanford Linear Collider, in which electrons and positrons are accelerated along a straight track 3.2 km/2 mi long and then steered into a head-on collision with other particles. The first linear accelerator was built in 1928 by the Norwegian engineer Ralph Wideröe to investigate the behaviour of heavy ions (large atoms with one or more electrons removed), but devices capable of accelerating smaller particles such as protons and electrons could not be built until after World War II and the development of high-power radio- and microwave-frequency generators.

liquefied petroleum gas (LPG)

liquid form of butane, propane, or pentane, produced by the distillation of petroleum during oil refining. At room temperature these substances are gases, although they can be easily liquefied and stored under pressure in metal containers. They are used for heating and cooking where other fuels are not available: camping stoves and cigarette lighters, for instance, often use liquefied butane as fuel.

liquid

state of matter between a solid and a gas. A liquid forms a level surface and assumes the shape of its container. Its atoms do not occupy fixed positions as in a crystalline solid, nor do they have freedom of movement as in a gas. Unlike a gas, a liquid is difficult to compress since pressure applied at one point is equally transmitted throughout (Pascal's principle). Hydraulics makes use of this property.

logic gate or logic circuit

one of the basic components used in building integrated circuits. The five basic types of gate make logical decisions based on the functions NOT, AND, OR, NAND (NOT AND), and NOR (NOT OR). With the exception of the NOT gate, each has two or more inputs.

The properties of a logic gate, or of a combination of gates, may be defined and presented in the form of a diagram called a *truth table*, which lists the output that will be triggered by each of the possible combinations of input signals. The process has close parallels in computer programming, where it forms the basis of binary logic.

US computer scientists announced the construction of logic gates from DNA in 1997. Rather than responding to an electronic signal their DNA gates respond to nucleotide sequences.

luminescence

emission of light from a body when its atoms are excited by means other than raising its temperature. Short-lived luminescence is called fluorescence; longer-lived luminescence is called phosphorescence.

When exposed to an external source of energy, the outer electrons in atoms of a luminescent substance absorb energy and 'jump' to a higher energy level. When these electrons 'jump' back to their former level they emit their excess energy as light. Many different exciting mechanisms are possible: visible light or other forms of electromagnetic radiation (ultraviolet rays or X-rays), electron bombardment, chemical reactions, friction, and radioactivity. Certain living organisms produce bioluminescence.

machine

device that allows a small force (the effort) to overcome a larger one (the load). There are three basic machines: the inclined plane (ramp), the lever, and the wheel and axle. All other machines are combinations of these three basic types. Simple machines derived from the inclined plane include the wedge, the gear, and the screw; the spanner is derived from the lever; the pulley from the wheel.

The principal features of a machine are its mechanical advantage, which is the ratio of load to effort, its velocity ratio, and its efficiency, which is the work done by the load divided by the work done by the effort; the latter is expressed as a percentage. In a perfect machine, with no friction, the efficiency would be 100%. All practical machines have efficiencies of less than 100%.

Mach number

ratio of the speed of a body to the speed of sound in the undisturbed medium through which the body travels. Mach 1 is reached when a body (such as an aircraft) has a velocity greater than that of sound ('passes the sound barrier'), namely 331 m/1,087 ft per second at sea level. It is named after Austrian physicist Ernst Mach (1838–1916).

magnet

any object that forms a magnetic field (displays magnetism), either permanently or temporarily through induction, causing it to attract materials such as iron, cobalt, nickel, and alloys of these. It always has two magnetic poles, called north and south.

The world's most powerful magnet was built in 1997 at the Lawrence Berkeley National Laboratory, California, USA. It produces a field 250,000 times stronger than the Earth's magnetic field (13.5 teslas). The coil magnet is made of an alloy of niobium and tin.

magnetic field

region around a permanent magnet, or around a conductor carrying an electric current, in which a force acts on a moving charge or on a magnet placed in the field. The field can be represented by lines of force parallel at each point to the direction of a small compass needle placed on them at that point. A magnetic field's magnitude is given by the magnetic flux density, expressed in teslas. See also polar reversal.

magnetic flux

measurement of the strength of the magnetic field around electric currents and magnets. Its SI unit is the weber; one weber per square metre is equal to one tesla.

The amount of magnetic flux through an area equals the product of the area and the magnetic field strength at a point within that area.

magnetic pole

region of a magnet in which its magnetic properties are strongest. Every magnet has two poles, called north and south. The north (or north-seeking) pole is so named because a freely suspended magnet will turn so that this pole points towards the Earth's magnetic north pole. The north pole of one magnet will be attracted to the south pole of another, but will be repelled by its north pole. So unlike poles attract, like poles repel.

magnetism

phenomenon associated with magnetic fields. Magnetic fields are produced by moving charged particles: in electromagnets, electrons flow through a coil of wire connected to a battery; in permanent magnets, spinning electrons within the atoms generate the field.

susceptibility

Substances differ in the extent to which they can be magnetized by an external field (susceptibility). Materials that can be strongly magnetized, such as iron, cobalt, and nickel, are said to be ferromagnetic; this is due to the formation of areas called domains in which atoms, weakly magnetic because of their spinning electrons, align to form areas of strong magnetism. Magnetic materials lose their magnetism if heated to the Curie temperature. Most other materials are paramagnetic, being only weakly pulled towards a strong magnet. This is because their atoms have a low level of magnetism and do not form domains. Diamagnetic materials are weakly repelled by a magnet since electrons within their atoms act as electromagnets and oppose the applied magnetic force. Antiferromagnetic materials have a very low susceptibility that increases with temperature; a similar phenomenon in materials such as ferrites is called ferrimagnetism.

application

Apart from its universal application in dynamos, electric motors, and switch gears, magnetism is of considerable importance in advanced technology – for example, in particle accelerators for nuclear research, memory stores for computers, tape recorders, and cryogenics.

magneton theory

Early in the 20th century, Pierre Weiss suggested the existence of the magneton or elementary magnet, an analogue of the electron, the elementary charge of electricity. This idea was developed by physicists, notably Albert Einstein, de Haas, and Niels Bohr. An electric current flowing round a circular coil has a magnetic field similar to that of a magnet whose axis coincides with that of the coil: the electrical theory of matter attempts to ascribe the magnetic properties of bodies to the orbital motions of the electrons in the atom. The quantum theory of the atom developed by Bohr supported the magneton theory, and subsequently direct experimental evidence of the existence of the magnetic moment associated with electron orbits was obtained by Otto Stern and Walther Gerlach in 1921.

magnetron

thermionic valve (electron tube) for generating very high-frequency oscillations, used in radar and to produce microwaves in a microwave oven. The flow of electrons from the tube's cathode to one or more anodes is controlled by an applied magnetic field.

maser acronym for microwave amplification by stimulated emission of radiation

high-frequency microwave amplifier or oscillator in which the signal to be amplified is used to stimulate excited atoms into emitting energy at the same frequency. Atoms or molecules are raised to a higher energy level and then allowed to lose this energy by radiation emitted at a precise frequency. The principle has been extended to other parts of the electromagnetic spectrum as, for example, in the laser.

The two-level ammonia-gas maser was first suggested in 1954 by US physicist Charles Townes at Columbia University, New York, USA, and independently the same year by Nikolai Basov and Aleksandr Prokhorov in Russia. The solid-state three-level maser, the most sensitive amplifier known, was envisaged by Nicolaas Bloembergen at Harvard, USA, in 1956. The ammonia maser is used as a frequency standard oscillator, and the three-level maser as a receiver for satellite communications and radio astronomy.

mass

quantity of matter in a body as measured by its inertia. Mass determines the acceleration produced in a body by a given force acting on it, the acceleration being inversely proportional to the mass of the body. The mass also determines the force exerted on a body by gravity on Earth, although this attraction varies slightly from place to place. In the SI system, the base unit of mass is the kilogram.

At a given place, equal masses experience equal gravitational forces, which are known as the weights of the bodies. Masses may, therefore, be compared by comparing the weights of bodies at the same place. The standard unit of mass to which all other masses are compared is a platinum-iridium cylinder of 1 kg, which is kept at the International Bureau of Weights and Measures in Sèvres, France.

mass–energy equation

German-born US physicist Albert Einstein's equation $E = mc^2$, denoting the equivalence of mass and energy. In SI units, E is the energy in joules, m is the mass in kilograms, and c, the speed of light in a vacuum, is in m per second.

mass spectrometer

apparatus for analysing chemical composition. Positive ions (charged particles) of a substance are separated by an electromagnetic system, designed to focus particles of equal mass to a point where they can be detected. This permits accurate measurement of the relative concentrations of the various ionic masses present, particularly isotopes.

matter

anything that has mass. All matter is made up of atoms, which in turn are made up of elementary particles; it ordinarily exists in one of three physical states: solid, liquid, or gas.

states of matter

Whether matter exists as a solid, liquid, or gas depends on its temperature and the pressure on it. Kinetic theory describes how the state of a material depends on the movement and arrangement of its atoms or molecules. In a solid the atoms or molecules vibrate in a fixed position. In a liquid, they do not occupy fixed positions as in a solid, and yet neither do they have the freedom of random movement that occurs within a gas, so the atoms or molecules within a liquid will always follow the shape of their container. The transition between states takes place at definite temperatures, called melting point and boiling point.

The history of science and philosophy is largely taken up with accounts of theories of matter, ranging from the hard atoms of Democritus to the 'proba-bility waves' of modern quantum theory. In the 19th century, investigations into the nature of combustion and chemical combination led to the firm identification of the atomic nature of the elements, and to the understanding that these atoms combined to form molecules, the smallest particles of chemical compounds (see atomic structure). The kinetic theory of matter, the theory of the motion of these atoms and molecules in gases, liquids, and solids, successfully accounted quantitatively for many of the phenomena of gases, liquids, and solids. In particular heat energy was shown to consist simply of the kinetic energy of the chaotic motion of the atoms and molecules in matter. Electrical research initiated by English physicist J J Thomson in the last decade of the 19th century heralded the spectacular progress in understanding of the 20th century. While the atom of English chemist John Dalton was the 'ultimate particle' in chemical reactions, it proved to be divisible by electrical means. Thomson identified the existence of the electron, a particle carrying a negative charge of electricity, and of mass nearly 2,000 times smaller than the lightest atom, hydrogen. Its mass and electrical charge were determined by US physicist Robert Millikan. Researches in radioactivity gradually established that in addition to negatively charged electrons, all matter is composed of positively charged protons, uncharged neutrons, and other shorter-lived particles.

Maxwell's screw rule
rule formulated by Scottish physicist James Maxwell that predicts the direction of the magnetic field produced around a wire carrying electric current. It states that if a right-handed screw is turned so that it moves forwards in the same direction as the conventional current, its direction of rotation will give the direction of the magnetic field.

mean free path
average distance travelled by a particle, atom, or molecule between successive collisions. It is of importance in the kinetic theory of gases.

mechanical advantage (MA)
ratio by which the load moved by a machine is greater than the effort applied to that machine. In equation terms: MA = load/effort.

The exact value of a working machine's MA is always less than its theoretical value because there will always be some frictional resistance that increases the effort necessary to do the work.

mechanical equivalent of heat
constant factor relating the calorie (the *c.g.s.* unit of heat) to the joule (the unit of mechanical energy), equal to 4.1868 joules per calorie. It is redundant in the SI system of units, which measures heat and all forms of energy in joules.

mechanics
branch of physics dealing with the motions of bodies and the forces causing these motions, and also with the forces acting on bodies in equilibrium. It is usually divided into dynamics and statics.

Quantum mechanics is the system based on the quantum theory, that has superseded Newtonian mechanics in the interpretation of physical phenomena on the atomic scale.

machines

As well as dealing with the direct action of forces on bodies, mechanics studies the nature and action of forces when they act on bodies by the agency of machinery. This gives the origin of the word 'mechanics': in its early stages it was the science of making machines. A machine in mechanics means any contrivance in which a force applied at one point is made to raise weight or overcome a resisting force acting at another point. All machines can be resolved into three primary machines: the lever, the inclined plane, and the wheel and axle.

The formulation of the theory of relativity by Albert Einstein led to a complete revision of the fundamentals of mechanics. When the bodies considered in a problem in mechanics have velocities (relative to a frame that is moving with a uniform or zero velocity with respect to the fixed stars) that are small compared with that of light, the old Newtonian or nonrelativistic mechanics still holds. But for larger velocities, important differences are predicted by relativity theory, and many have been confirmed by experiment.

melting point

temperature at which a substance melts, or changes from solid to liquid form. A pure substance under standard conditions of pressure (usually one atmosphere) has a definite melting point. If heat is supplied to a solid at its melting point, the temperature does not change until the melting process is complete. The melting point of ice is 0°C or 32°F.

meson

group of unstable subatomic particles made up of a quark and an antiquark. It is found in cosmic radiation, and is emitted by nuclei under bombardment by very high-energy particles.

The mesons form a subclass of the hadrons and include the kaons and pions. Their existence was predicted in 1935 by Japanese physicist Hideki Yukawa.

moderator

material such as graphite or heavy water used to reduce the speed of high-energy neutrons in a nuclear reactor. Neutrons produced by nuclear fission are fast-moving and must be slowed to initiate further fission so that nuclear energy continues to be released at a controlled rate.

Slow neutrons are much more likely to cause fission in a uranium-235 nucleus than to be captured in a U-238 (nonfissile uranium) nucleus. By using a moderator, a reactor can thus be made to work with fuel containing only a small proportion of U-235.

modulation

variation of frequency, or amplitude, of a radio carrier wave, in accordance with the audio characteristics of the speaking voice, music, or other signal being transmitted.

See also pulse-code modulation, amplitude modulation (AM), and frequency modulation (FM).

molecule

molecules are the smallest particles of an element or compound that can exist independently. Hydrogen atoms, at room temperature, do not exist independently. They are bonded in pairs to form hydrogen molecules. A molecule of a compound consists of two or more different atoms bonded together. Molecules vary in size and complexity from the hydrogen molecule (H_2) to the large macromolecules of proteins. They may be held together by ionic bonds, in which the atoms gain or lose electrons to form ions, or by covalent bonds, where electrons from each atom are shared in a new molecular orbital. Each compound is represented by a chemical symbol, indicating the elements into which it can be broken down and the number of each type of atom present. The symbolic representation of a molecule is known as its formula. For example, one molecule of the compound water, having two atoms of hydrogen and one atom of oxygen, is shown as H_2O.

moment of a force

measure of the turning effect, or torque, produced by a force acting on a body. It is equal to the product of the force and the perpendicular distance from its line of action to the point, or pivot, about which the body will turn. Its unit is the newton metre.

If the magnitude of the force is F newtons and the perpendicular distance is d metres then the moment is given by:

moment $= Fd$

moment of inertia

sum of all the point masses of a rotating object multiplied by the squares of their respective distances from the axis of rotation.

In linear dynamics, Newton's second law of motion states that the force F on a moving object equals the products of its mass m and acceleration a ($F = ma$); the analogous equation in rotational dynamics is $T = I\alpha$, where T is the torque (the turning effect of a force) that causes an angular acceleration α and I is the moment of inertia. For a given object, I depends on its shape and the position of its axis of rotation.

momentum

product of the mass of a body and its velocity. If the mass of a body is m kilograms and its velocity is v m s^{-1}, then its momentum is given by:

momentum = mv

Its unit is the kilogram metre-per-second (kg m s⁻¹) or the newton second. The momentum of a body does not change unless a resultant or unbalanced force acts on that body (see Newton's laws of motion).

According to Newton's second law of motion, the magnitude of a resultant force F equals the rate of change of momentum brought about by its action, or:

$$F = \frac{(mv - mu)}{t}$$

where mu is the initial momentum of the body, mv is its final momentum, and t is the time in seconds over which the force acts. The change in momentum, or impulse, produced can therefore be expressed as:

impulse = $mv - mu = Ft$

The law of conservation of momentum is one of the fundamental concepts of classical physics. It states that the total momentum of all bodies in a closed system is constant and unaffected by processes occurring within the system. The **angular momentum** of an orbiting or rotating body of mass m travelling at a velocity v in a circular orbit of radius R is expressed as mvR. Angular momentum is conserved, and should any of the values alter (such as the radius of orbit), the other values (such as the velocity) will compensate to preserve the value of angular momentum, and that lost by one component is passed to another.

motor effect

tendency of a wire carrying an electric current in a magnetic field to move. The direction of the movement is given by the left-hand rule (see Fleming's rules). This effect is used in the electric motor. It also explains why streams of electrons produced, for instance, in a television tube can be directed by electromagnets.

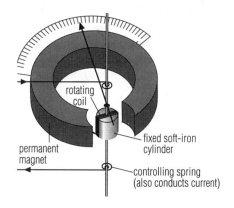

moving-coil meter

instrument used to detect and measure electrical current. A coil of wire pivoted between the poles of a permanent magnet is turned by the motor effect of an electric current (by which a force acts on a wire carrying a current in a magnetic

A simple moving-coil meter. Direct electric current (DC) flowing through the wire coil combined with the presence of a magnetic field causes the coil to rotate; this in turn moves a pointer across a calibrated scale so that the degree of rotation can be related to the magnitude of the current.

field). The extent to which the coil turns can then be related to the magnitude of the current.

The sensitivity of the instrument depends directly upon the strength of the permanent magnet used, the number of turns making up the moving coil, and the coil's area. It depends inversely upon the strength of the controlling springs used to restrain the rotation of the coil. By the addition of a suitable resistor, a moving-coil meter can be adapted to read potential difference in volts.

muon
elementary particle similar to the electron except for its mass which is 207 times greater than that of the electron. It has a half-life of 2 millionths of a second, decaying into electrons and neutrinos. The muon was originally thought to be a meson but is now classified as a lepton. See also tau.

natural frequency
frequency at which a mechanical system will vibrate freely. A pendulum, for example, always oscillates at the same frequency when set in motion. More complicated systems, such as bridges, also vibrate with a fixed natural frequency. If a varying force with a frequency equal to the natural frequency is applied to such an object the vibrations can become violent, a phenomenon known as resonance.

neutrino
any of three uncharged elementary particles (and their antiparticles) of the lepton class, having a mass that is very small (possibly zero). The most familiar type, the antiparticle of the electron neutrino, is emitted in the beta decay of a nucleus. The other two are the muon and tau neutrinos.

Supernova 1987A was the first object outside the Solar System to be observed by neutrino emission. The Sun emits neutrinos, but in smaller numbers than theoretically expected. The shortage of solar neutrinos is one of the biggest mysteries in modern astrophysics.

neutron
one of the three main subatomic particles, the others being the proton and the electron. The neutron is a composite particle, being made up of three quarks, and therefore belongs to the baryon group of the hadrons. Neutrons have about the same mass as protons but no electric charge, and occur in the nuclei of all atoms except hydrogen. They contribute to the mass of atoms but do not affect their chemistry.

For instance, the isotopes of a single element differ only in the number of neutrons in their nuclei but have identical chemical properties. Outside a nucleus, a free neutron is unstable, decaying with a half-life of 11.6 minutes into a proton, an electron, and an antineutrino. The neutron was discovered by the British chemist James Chadwick in 1932. Neutrons and protons have masses approximately 2,000 times those of electrons. The process by which a neutron changes into a proton is called beta decay.

newton

SI unit (symbol N) of force. One newton is the force needed to accelerate an object with mass of one kilogram by one metre per second per second. The weight of a medium size (100 g/3 oz) apple is one newton.

Newton's laws of motion

three laws that form the basis of Newtonian mechanics. (1) Unless acted upon by an unbalanced force, a body at rest stays at rest, and a moving body continues moving at the same speed in the same straight line. (2) An unbalanced force applied to a body gives it an acceleration proportional to the force (and in the direction of the force) and inversely proportional to the mass of the body. (3) When a body A exerts a force on a body B, B exerts an equal and opposite force on A; that is, to every action there is an equal and opposite reaction.

nuclear energy or atomic energy

energy released from the inner core, or nucleus, of the atom. Energy produced by nuclear fission (the splitting of uranium or plutonium nuclei) has been

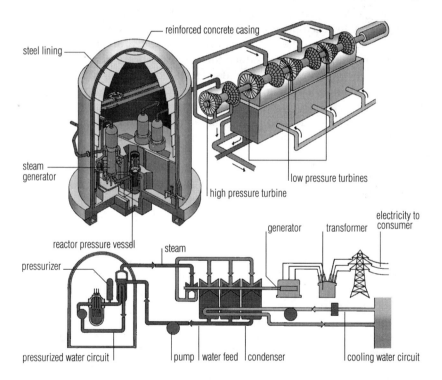

A pressurized water nuclear power station. Water at high pressure is circulated around the reactor vessel where it is heated. The hot water is pumped to the steam generator where it boils in a separate circuit; the steam drives the turbines coupled to the electricity generator. This is the most widely used type of reactor. More than 20 countries have pressurized water reactors.

harnessed since the 1950s to generate electricity, and research continues into the possible controlled use of nuclear fusion (the fusing, or combining, of atomic nuclei).

In nuclear power stations, fission takes place in a nuclear reactor. The nuclei of uranium or, more rarely, plutonium are induced to split, releasing large amounts of heat energy. The heat is then removed from the core of the reactor by circulating gas or water, and used to produce the steam that drives alternators and turbines to generate electrical power. Unlike fossil fuels, such as coal and oil, which must be burned in large quantities to produce energy, nuclear fuels are used in very small amounts and supplies are therefore unlikely to be exhausted in the foreseeable future. However, the use of nuclear energy has given rise to concern over safety. Anxiety has been heightened by accidents such as the one at Chernobyl, Ukraine, in 1986. There has also been mounting concern about the production and disposal of toxic nuclear waste, which may have an active life of several thousand years, and the cost of maintaining nuclear power stations and decommissioning them at the end of their lives.

nuclear fusion
process whereby two atomic nuclei are fused, with the release of a large amount of energy. Very high temperatures and pressures are required for the process. Under these conditions the atoms involved are stripped of all their electrons so that the remaining particles, which together make up a *plasma*, can come close together at very high speeds and overcome the mutual repulsion of the positive charges on the atomic nuclei. At very close range the strong nuclear force will come into play, fusing the particles to form a larger nucleus. As fusion is accompanied by the release of large amounts of energy, the process might one day be harnessed to form the basis of commercial energy production. Methods of achieving controlled fusion are therefore the subject of research around the world.

Fusion is the process by which the Sun and the other stars produce their energy.

nuclear physics
study of the properties of the nucleus of the atom, including the structure of nuclei; nuclear forces; the interactions between particles and nuclei; and the study of radioactive decay. The study of elementary particles is particle physics.

nuclear reaction
reaction involving the nuclei of atoms. Atomic nuclei can undergo changes either as a result of radioactive decay, as in the decay of radium to radon (with the emission of an alpha particle):

$$^{226}_{88}\text{Ra} \rightarrow {}^{222}_{86}\text{Rn} + {}^{4}_{2}\text{He}$$

or as a result of particle bombardment in a machine or device, as in the production of cobalt-60 by the bombardment of cobalt-59 with neutrons:

$$^{59}_{27}\text{Co} + {}^{1}_{0}\text{n} \rightarrow {}^{60}_{27}\text{Co} + \gamma$$

Nuclear fission and nuclear fusion are examples of nuclear reactions. The enormous amounts of energy released arise from the mass–energy relation put forward by Einstein, stating that $E = mc^2$ (where E is energy, m is mass, and c is the velocity of light).

In nuclear reactions the sum of the masses of all the products is less than the sum of the masses of the reacting particles. This lost mass is converted to energy according to Einstein's equation.

nucleon
either a proton or a neutron, when present in the atomic nucleus. *Nucleon number* is an alternative name for the mass number of an atom.

nucleus
positively charged central part of an atom, which constitutes almost all its mass. Except for hydrogen nuclei, which have only protons, nuclei are composed of both protons and neutrons. Surrounding the nuclei are electrons, of equal and opposite charge to that of the protons, thus giving the atom a neutral charge.

The nucleus was discovered by New Zealand-born British physicist Ernest Rutherford in 1911 as a result of experiments in passing alpha particles through very thin gold foil.

A few of the particles were deflected back, and Rutherford, astonished, reported: 'It was almost as if you fired a 15-inch shell at a piece of tissue paper and it came back and hit you!' The deflection, he deduced, was due to the positively charged alpha particles being repelled by approaching a small but dense positively charged nucleus.

ohm
SI unit (symbol Ω) of electrical resistance (the property of a conductor that restricts the flow of electrons through it).

It was originally defined with reference to the resistance of a column of mercury, but is now taken as the resistance between two points when a potential difference of one volt between them produces a current of one ampere.

Ohm's law
law that states that, for many materials over a wide range of conditions, the current flowing in a conductor maintained at constant temperature is directly proportional to the potential difference (voltage) between its ends. The law was discovered by German physicist Georg Ohm in 1827.

If a current of I amperes flows between two points in a conductor across which the potential difference is V volts, then V/I is a constant called the resistance R ohms between those two points. Hence: $V/I = R$ or $V = IR$ Not all conductors obey Ohm's law; those that do are called *ohmic conductors*.

operational amplifier or op-amp
electronic circuit that is used as a basic building block in electronic design. Operational amplifiers are used in a wide range of electronic measuring

instruments. The name arose because they were originally designed to carry out mathematical operations and solve equations.

The voltage *gain* of an inverting operational amplifier is equal to the ratio of the resistance of the feedback resistor to the resistance of the input resistor.

optics
branch of physics that deals with the study of light and vision – for example, shadows and mirror images, lenses, microscopes, telescopes, and cameras. On striking a surface, light rays are reflected or refracted with some absorption of energy, and the study of this is known as geometrical optics.

orbital, atomic
region around the nucleus of an atom (or, in a molecule, around several nuclei) in which an electron is likely to be found. According to quantum theory, the position of an electron is uncertain; it may be found at any point. However, it is more likely to be found in some places than in others, and this pattern of probabilities makes up the orbital.

An atom or molecule has numerous orbitals, each of which has a fixed size and shape. An orbital is characterized by three numbers, called quantum numbers, representing its energy (and hence size), its angular momentum (and hence shape), and its orientation. Each orbital can be occupied by one or (if their spins are aligned in opposite directions) two electrons.

oscillation
one complete to-and-fro movement of a vibrating object or system. For any particular vibration, the time for one oscillation is called its period and the number of oscillations in one second is called its frequency. The maximum displacement of the vibrating object from its rest position is called the amplitude of the oscillation.

oscilloscope
another name for *cathode-ray oscilloscope.*

parallelogram of forces
method of calculating the resultant (combined effect) of two different forces acting together on an object. Because a force has both magnitude and direction it is a vector quantity and can be represented by a straight line. A second force acting at the same point in a different direction can be represented by another line drawn at an angle to the first. By completing the parallelogram (of which the two lines are sides) a diagonal may be drawn from the original angle to the opposite corner to represent the resultant force vector.

particle detector
one of a number of instruments designed to detect subatomic particles and track their paths; they include the cloud chamber, bubble chamber, spark chamber, and multiwire chamber.

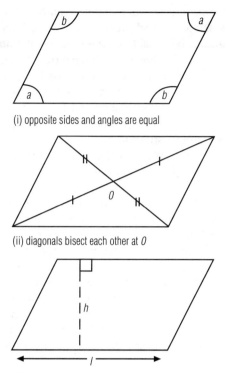

(i) opposite sides and angles are equal

(ii) diagonals bisect each other at *O*

(iii) area of a parallelogram *l* x *h*

The diagram shows how the parallelogram of forces can be used to calculate the resultant (combined effect) of two different forces acting together on an object. The two forces are represented by two lines drawn at an angle to each other. By completing the parallelogram (of which the two lines are sides), a diagonal may be drawn from the original angle to the opposite corner to represent the resultant force vector.

The earliest particle detector was the cloud chamber, which contains a super-saturated vapour in which particles leave a trail of droplets, in much the same way that a jet aircraft leaves a trail of vapour in the sky. A bubble chamber contains a superheated liquid in which a particle leaves a trail of bubbles. A spark chamber contains a series of closely-packed parallel metal plates, each at a high voltage. As particles pass through the chamber, they leave a visible spark between the plates. A modern multiwire chamber consists of an array of fine, closely-packed wires, each at a high voltage. As a particle passes through the chamber, it produces an electrical signal in the wires. A computer analyses the signal and reconstructs the path of the particles. Multiwire detectors can be used to detect X-ray and gamma rays, and are used as detectors in positron emission tomography (PET).

particle physics
study of the particles that make up all atoms, and of their interactions. More than 300 subatomic particles have now been identified by physicists, categorized into several classes according to their mass, electric charge, spin,

magnetic moment, and interaction. Subatomic particles include the elementary particles (quarks, leptons, and gauge bosons), which are indivisible, so far as is known, and so may be considered the fundamental units of matter; and the hadrons (baryons, such as the proton and neutron, and mesons), which are composite particles, made up of two or three quarks. Quarks, protons, electrons, and neutrinos are the only stable particles (the neutron being stable only when in the atomic nucleus). The unstable particles decay rapidly into other particles, and are known from experiments with particle accelerators and cosmic radiation. See atomic structure.

Pioneering research took place at the Cavendish Laboratory, Cambridge University, UK. In 1897 the English physicist J J Thomson discovered that all atoms contain identical, negatively charged particles (electrons), which can easily be freed. By 1911 the New Zealand physicist Ernest Rutherford had shown that the electrons surround a very small, positively-charged nucleus. In the case of hydrogen, this was found to consist of a single positively charged particle, a proton. The nuclei of other elements are made up of protons and uncharged particles called neutrons. 1932 saw the discovery of a particle (whose existence had been predicted by the British theoretical physicist Paul Dirac in 1928) with the mass of an electron, but an equal and opposite charge – the positron. This was the first example of antimatter; it is now believed that all particles have corresponding antiparticles. In 1934 the Italian-born US physicist Enrico Fermi argued that a hitherto unsuspected particle, the neutrino, must accompany electrons in beta-emission.

particles and fundamental forces
By the mid-1930s, four types of fundamental force interacting between particles had been identified. The electromagnetic force (1) acts between all particles with electric charge, and is related to the exchange between these particles of gauge bosons called photons, packets of electromagnetic radiation. In 1935 the Japanese physicist Hideki Yukawa suggested that the strong nuclear force (2) (binding protons and neutrons together in the nucleus) was transmitted by the exchange of particles with a mass about one-tenth of that of a proton; these particles, called pions (originally pi mesons), were found by the British physicist Cecil Powell in 1946. Yukawa's theory was largely superseded from 1973 by the theory of quantum chromodynamics, which postulates that the strong nuclear force is transmitted by the exchange of gauge bosons called gluons between the quarks and antiquarks making up protons and neutrons. Theoretical work on the weak nuclear force (3) began with Enrico Fermi in the 1930s. The existence of the gauge bosons that carry this force, the W and Z particles, was confirmed in 1983 at CERN, the European nuclear research organization. The fourth fundamental force, gravity (4), is experienced by all matter; the postulated carrier of this force has been named the graviton.

leptons
The electron, muon, tau, and their neutrinos comprise the leptons – particles with half-integral spin that 'feel' the weak nuclear and electromagnetic force but not the strong force. The muon (found by the US physicist Carl Anderson

in cosmic radiation in 1937) produces the muon neutrino when it decays; the tau, a surprise discovery of the 1970s, produces the tau neutrino when it decays.

mesons and baryons
The hadrons (particles that 'feel' the strong nuclear force) were found in the 1950s and 1960s. They are classified into mesons, with whole-number or zero spins, and baryons (which include protons and neutrons), with half-integral spins. It was shown in the early 1960s that if hadrons of the same spin are represented as points on suitable charts, simple patterns are formed. This symmetry enabled a hitherto unknown baryon, the omega-minus, to be predicted from a gap in one of the patterns; it duly turned up in experiments.

quarks
In 1964 the US physicists Murray Gell-Mann and George Zweig suggested that all hadrons were built from three 'flavours' of a new particle with half-integral spin and a charge of magnitude either $-\frac{1}{3}$ or $+\frac{2}{3}$ that of an electron; Gell-Mann named the particle the quark. Mesons are quark–antiquark pairs (spins either add to one or cancel to zero), and baryons are quark triplets. To account for new mesons such as the psi (J) particle the number of quark flavours had risen to six by 1985.

particle, subatomic
particle that is smaller than an atom; see particle physics.

Pauli exclusion principle
principle of atomic structure. See exclusion principle.

Peltier effect
change in temperature at the junction of two different metals produced when an electric current flows through them. The extent of the change depends on what the conducting metals are, and the nature of change (rise or fall in temperature) depends on the direction of current flow. It is the reverse of the Seebeck effect. It is named after the French physicist Jean Charles Peltier (1785–1845) who discovered it in 1834.

pendulum
weight (called a 'bob') swinging at the end of a rod or cord. The regularity of a pendulum's swing was used in making the first really accurate clocks in the 17th century. Pendulums can be used for measuring the acceleration due to gravity (an important constant in physics).

Specialized pendulums are used to measure velocities (ballistic pendulum) and to demonstrate the Earth's rotation (Foucault's pendulum).

perpetual motion
idea that a machine can be designed and constructed in such a way that, once started, it will do work indefinitely without requiring any further input of energy

(motive power). Such a device would contradict at least one of the two laws of thermodynamics that state that (1) energy can neither be created nor destroyed (the law of conservation of energy) and (2) heat cannot by itself flow from a cooler to a hotter object. As a result, all practical (real) machines require a continuous supply of energy, and no heat engine is able to convert all the heat into useful work.

phase

stage in an oscillatory motion, such as a wave motion: two waves are in phase when their peaks and their troughs coincide. Otherwise, there is a *phase difference*, which has consequences in interference phenomena and alternating current electricity.

photocell or photoelectric cell

device for measuring or detecting light or other electromagnetic radiation, since its electrical state is altered by the effect of light. In a *photoemissive* cell, the radiation causes electrons to be emitted and a current to flow (photoelectric effect); a *photovoltaic* cell causes an electromotive force to be generated in the presence of light across the boundary of two substances. A *photoconductive* cell, which contains a semiconductor, increases its conductivity when exposed to electromagnetic radiation.

photodiode

semiconductor p–n junction diode used to detect light or measure its intensity. The photodiode is encapsulated in a transparent plastic case that allows light to fall onto the junction. When this occurs, the reverse-bias resistance (high resistance in the opposite direction to normal current-flow) drops and allows a larger reverse-biased current to flow through the device. The increase in current can then be related to the amount of light falling on the junction.

Photodiodes that can detect small changes in light level are used in alarm systems, camera exposure controls, and optical communication links.

photoelectric cell

alternative name for *photocell.*

photon

elementary particle or 'package' (quantum) of energy in which light and other forms of electromagnetic radiation are emitted. The photon has both particle and wave properties; it has no charge, is considered massless but possesses momentum and energy. It is one of the gauge bosons, and is the carrier of the electromagnetic force, one of the fundamental forces of nature.

According to quantum theory the energy of a photon is given by the formula $E = hf$, where h is Planck's constant and f is the frequency of the radiation emitted.

piezoelectric effect
property of some crystals (for example, quartz) to develop an electromotive force or voltage across opposite faces when subjected to tension or compression, and, conversely, to expand or contract in size when subjected to an electromotive force. Piezoelectric crystal oscillators are used as frequency standards (for example, replacing balance wheels in watches), and for producing ultrasound.

pion or pi meson
subatomic particle with a neutral form (mass 135 MeV) and a charged form (mass 139 MeV). The charged pion decays into muons and neutrinos and the neutral form decays into gamma-ray photons. They belong to the hadron class of elementary particles.

The mass of a charged pion is 273 times that of an electron; the mass of a neutral pion is 264 times that of an electron.

pitch
distance between the adjacent threads of a screw or bolt. When a screw is turned through one full turn it moves a distance equal to the pitch of its thread. A screw thread is a simple type of machine, acting like a rolled-up inclined plane, or ramp (as may be illustrated by rolling a long paper triangle around a pencil). A screw has a mechanical advantage greater than one.

Planck's constant
fundamental constant (symbol h) that relates the energy (E) of one quantum of electromagnetic radiation (a 'packet' of energy; see quantum theory) to the frequency (f) of its radiation by $E = hf$. Its value is 6.6262×10^{-34} joule seconds.

plasma
ionized gas produced at extremely high temperatures, as in the Sun and other stars. It contains positive and negative charges in equal numbers. It is a good electrical conductor. In thermonuclear reactions the plasma produced is confined through the use of magnetic fields.

polarized light
light in which the electromagnetic vibrations take place in one particular plane. In ordinary (unpolarized) light, the electric fields vibrate in all planes perpendicular to the direction of propagation. After reflection from a polished surface or transmission through certain materials (such as Polaroid), the electric fields are confined to one direction, and the light is said to be *linearly polarized*. In *circularly polarized* and *elliptically polarized* light, the electric fields are confined to one direction, but the direction rotates as the light propagates. Polarized light is used to test the strength of sugar solutions and to measure stresses in transparent materials.

positron

antiparticle of the electron; an elementary particle having the same mass as an electron but exhibiting a positive charge. The positron was discovered in 1932 by US physicist Carl Anderson at the California Institute of Technology, USA, its existence having been predicted by the British physicist Paul Dirac in 1928.

positron emission tomography (PET)

imaging technique that enables doctors to observe the metabolic activity of the human body by following the progress of a radioactive chemical that has been inhaled or injected, detecting gamma radiation given out when positrons emitted by the chemical are annihilated. The technique has been used to study a wide range of conditions, including schizophrenia, Alzheimer's disease, and Parkinson's disease.

potential difference (PD)

difference in the electrical potential (see potential, electric) of two points, being equal to the electrical energy converted by a unit electric charge moving from one point to the other. The SI unit of potential difference is the volt (V). The potential difference between two points in a circuit is commonly referred to as voltage. See also Ohm's law.

In equation terms, potential difference V may be defined by:

$$V = W/Q$$

here W is the electrical energy converted in joules and Q is the charge in coulombs. The unit of potential difference is the volt.

potential, electric

energy required to bring a unit electric charge from infinity to the point at which potential is defined. The SI unit of potential is the volt (V). Positive electric charges will flow 'downhill' from a region of high potential to a region of low potential.

A charged conductor, for example, has a higher potential than the Earth, the potential of which is taken by convention to be zero. An electric cell (battery) has a potential difference between its terminals, which can make current flow in an external circuit. The difference in potential – potential difference (pd) – is expressed in volts so, for example, a 12 V battery has a pd of 12 volts between its negative and positive terminals.

potential energy (PE)

energy possessed by an object by virtue of its relative position or state (for example, as in a compressed spring or a muscle). It is contrasted with kinetic energy, the form of energy possessed by moving bodies. An object that has been raised up is described as having gravitational potential energy.

power

measure of the amount by which a lens will deviate light rays. A powerful converging lens will converge parallel rays strongly, bringing them to a focus at a short distance from the lens. The unit of power is the *dioptre*, which is equal to the reciprocal of focal length in metres. By convention, the power of a converging (or convex) lens is positive and that of a diverging (or concave) lens negative.

power

rate of doing work or consuming energy. It is measured in watts (joules per second) or other units of work per unit time.

If the work done or energy consumed is *W* joules and the time taken is *t* seconds, then the power *P* is given by the formula:

$$P = \frac{W}{t}$$

power station

building where electrical energy is generated from a fuel or from another form of energy. Fuels used include fossil fuels such as coal, gas, and oil, and the nuclear fuel uranium. Renewable sources of energy include gravitational potential energy, used to produce hydroelectric power, and wind power.

Large coal-fired power stations such as this one at Didcot, southern England, require huge amounts of coal to operate – Didcot consumes up to 1,000 truckfuls per day. The electricity produced per tonne of coal may be as high as 2,000 kwh. Concern about acid rain and global warming, however, has led to a move away from coal-fired power generation. AEA Technology

The energy supply is used to turn turbines either directly by means of water or wind pressure, or indirectly by steam pressure, steam being generated by burning fossil fuels or from the heat released by the fission of uranium nuclei. The turbines in their turn spin alternators, which generate electricity at very high voltage.

pressure
force that would act normally (at right angles) per unit surface area of a body immersed in the fluid. The SI unit of pressure is the pascal (Pa), equal to a pressure of one newton per square metre. In the atmosphere, the pressure declines with height from about 100 kPa at sea level to zero where the atmosphere fades into space. Pressure is commonly measured with a barometer, manometer, or Bourdon gauge. Other common units of pressure are the bar and the torr.

Absolute pressure is measured from a vacuum; *gauge pressure* is the difference between the absolute pressure and the local atmospheric pressure. In a liquid, the pressure at a depth h is given by ρgh where ρ is the density and g is the acceleration of free fall.

proton
positively charged subatomic particle, a constituent of the nucleus of all atoms. It belongs to the baryon subclass of the hadrons. A proton is extremely long-lived, with a lifespan of at least 10^{32} years. It carries a unit positive charge equal to the negative charge of an electron. Its mass is almost 1,836 times that of an electron, or 1.67×10^{-27} kg. Protons are composed of two up quarks and one down quark held together by gluons. The number of protons in the atom of an element is equal to the atomic number of that element.

pulse-code modulation (PCM)
form of digital modulation in which microwaves or light waves (the carrier waves) are switched on and off in pulses of varying length according to a binary code. It is a relatively simple matter to transmit data that are already in binary code, such as those used by computer, by these means. However, if an analogue audio signal is to be transmitted, it must first be converted to a *pulse-amplitude modulated* signal (PAM) by regular sampling of its amplitude. The value of the amplitude is then converted into a binary code for transmission on the carrier wave.

quantum chromodynamics (QCD)
theory describing the interactions of quarks, the elementary particles that make up all hadrons (subatomic particles such as protons and neutrons). In quantum chromodynamics, quarks are considered to interact by exchanging particles called gluons, which carry the strong nuclear force, and whose role is to 'glue' quarks together.

The mathematics involved in the theory is complex, and, although a number of successful predictions have been made, the theory does not compare in accuracy with quantum electrodynamics, upon which it is modelled. See forces, fundamental.

quantum electrodynamics (QED)

theory describing the interaction of charged subatomic particles within electric and magnetic fields. It combines quantum theory and relativity, and considers charged particles to interact by the exchange of photons. QED is remarkable for the accuracy of its predictions; for example, it has been used to calculate the value of some physical quantities to an accuracy of ten decimal places, a feat equivalent to calculating the distance between New York and Los Angeles to within the thickness of a hair. The theory was developed by US physicists Richard Feynman and Julian Schwinger and by Japanese physicist Sin-Itiro Tomonaga in 1948.

quantum number

one of a set of four numbers that uniquely characterize an electron and its state in an atom. The *principal quantum number n* defines the electron's main energy level. The *orbital quantum number l* relates to its angular momentum. The *magnetic quantum number m* describes the energies of electrons in a magnetic field. The *spin quantum number m_s* gives the spin direction of the electron.

The principal quantum number, defining the electron's energy level, corresponds to shells (energy levels) also known by their spectroscopic designations K, L, M, and so on. The orbital quantum number gives rise to a series of subshells designated *s, p, d, f,* and so on, of slightly different energy levels. The magnetic quantum number allows further subdivision of the subshells (making three subdivisions p_x, p_y, and p_z in the *p* subshell, for example, of the same energy level). No two electrons in an atom can have the same set of quantum numbers (the Pauli exclusion principle).

quantum theory or quantum mechanics

theory that energy does not have a continuous range of values, but is, instead, absorbed or radiated discontinuously, in multiples of definite, indivisible units called quanta. Just as earlier theory showed how light, generally seen as a wave motion, could also in some ways be seen as composed of discrete particles (photons), quantum theory shows how atomic particles such as electrons may also be seen as having wavelike properties. Quantum theory is the basis of particle physics, modern theoretical chemistry, and the solid-state physics that describes the behaviour of the silicon chips used in computers.

quark

elementary particle that is the fundamental constituent of all hadrons (subatomic particles that experience the strong nuclear force and divided into baryons, such as neutrons and protons, and mesons). Quarks have electric charges that are fractions of the electronic charge $+\frac{2}{3}$ or $-\frac{1}{3}$ of the electronic charge). There are six types, or 'flavours': up, down, top, bottom, strange, and charmed, each of which has three varieties, or 'colours': red, green, and blue (visual colour is not meant, although the analogy is useful in many ways). To each quark there is an antiparticle, called an antiquark. See quantum chromodynamics (QCD).

The existence of the top quark was confirmed by two teams of physicists at Fermilab in March 1995. It is unstable and unlikely to last more than a millionth of a billionth of a billionth of a second, but experiments at Fermilab suggest it is at least as massive as a silver atom, and 20,000 times heavier than an up quark.

radar acronym for radio direction and ranging
device for locating objects in space, direction finding, and navigation by means of transmitted and reflected high-frequency radio waves.

The direction of an object is ascertained by transmitting a beam of short-wavelength (1–100 cm/0.5–40 in), short-pulse radio waves, and picking up the reflected beam. Distance is determined by timing the journey of the radio waves (travelling at the speed of light) to the object and back again. Radar is also used to detect objects underground, for example service pipes, and in archaeology. Contours of remains of ancient buildings can be detected down to 20 m/66 ft below ground. Radar is essential to navigation in darkness, cloud, and fog, and is widely used in warfare to detect enemy aircraft and missiles. To avoid detection, various devices, such as modified shapes (to reduce their radar cross-section), radar absorbent paints and electronic jamming are used. To pinpoint small targets laser 'radar', instead of microwaves, has been developed. Developed independently in the UK, France, Germany, and the USA in the 1930s, it was first put to practical use for aircraft detection by the British, who had a complete coastal chain of radar stations installed by September 1938. Radar is also used in meteorology and astronomy.

radiation
emission of radiant energy as particles or waves – for example, heat, light, alpha particles, and beta particles (see electromagnetic waves and radioactivity). See also atomic radiation.

Of the radiation given off by the Sun, only a tiny fraction of it, called insolation, reaches the Earth's surface; much of it is absorbed and scattered as it passes through the atmosphere. The radiation given off by the Earth itself is called *ground radiation*.

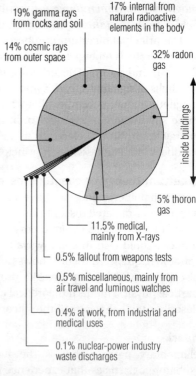

Pie chart showing the various sources of radiation in the environment. Most radiation is from natural sources, such as radioactive minerals, but 13% comes from the by-products of human activities.

radiation units

units of measurement for radioactivity and radiation doses. In SI units, the activity of a radioactive source is measured in becquerels (symbol Bq), where one becquerel is equal to one nuclear disintegration per second (an older unit is the curie). The exposure is measured in coulombs per kilogram (C kg^{-1}); the amount of ionizing radiation (X-rays or gamma rays) which produces one coulomb of charge in one kilogram of dry air (replacing the roentgen). The absorbed dose of ionizing radiation is measured in grays (symbol Gy) where one gray is equal to one joule of energy being imparted to one kilogram of matter (the rad is the previously used unit). The dose equivalent, which is a measure of the effects of radiation on living organisms, is the absorbed dose multiplied by a suitable factor that depends upon the type of radiation. It is measured in sieverts (symbol Sv), where one sievert is a dose equivalent of one joule per kilogram (an older unit is the rem).

radioactive decay

process of disintegration undergone by the nuclei of radioactive elements, such as radium and various isotopes of uranium and the transuranic elements. This changes the element's atomic number, thus transmuting one element into another, and is accompanied by the emission of radiation. Alpha and beta decay are the most common forms.

In *alpha decay* (the loss of a helium nucleus – two protons and two neutrons) the atomic number decreases by two and a new nucleus is formed, for example, an atom of uranium isotope of mass 238, on emitting an alpha particle, becomes an atom of thorium, mass 234. In *beta decay* the loss of an electron from an atom is accomplished by the transformation of a neutron into a proton, thus resulting in an increase in the atomic number of one. For example, the decay of the carbon-14 isotope results in the formation of an atom of nitrogen (mass 14, atomic number 7) and the emission of an electron. Gamma emission usually occurs as part of alpha or beta emission. In gamma emission high-speed electromagnetic radiation is emitted from the nucleus, making it more stable during the loss of an alpha or beta particle. Certain lighter artificially created isotopes also undergo radioactive decay. The associated radiation consists of alpha rays, beta rays, or gamma rays (or a combination of these), and it takes place at a constant rate expressed as a specific half-life, which is the time taken for half of any mass of that particular isotope to decay completely. Less commonly occurring decay forms include heavy-ion emission, electron capture, and spontaneous fission (in each of these the atomic number decreases). The original nuclide is known as the parent substance, and the product is a daughter nuclide (which may or may not be radioactive). The final product in all modes of decay is a stable element.

radioactivity

spontaneous alteration of the nuclei of radioactive atoms, accompanied by the emission of radiation. It is the property exhibited by the radioactive isotopes of stable elements and all isotopes of radioactive elements, and can be either natural or induced. See radioactive decay.

the discovery of radioactivity

Radioactivity was first discovered in 1896, when French physicist Henri Becquerel observed that some photographic plates, although securely wrapped up, became blackened when placed near certain uranium compounds. A closer investigation showed that thin metal coverings were unable to prevent the blackening of the plates. It was clear that the uranium compounds emitted radiation that was able to penetrate the metal coverings. Pierre and Marie Curie soon succeeded in isolating other radioactive elements. One of these was radium, which was found to be over 1 million times more radioactive than uranium.

radioactive radiations

Ernest Rutherford revealed that there are three types of radiation: alpha particles, beta particles, and gamma rays. Alpha particles are positively charged, high-energy particles emitted from the nucleus of a radioactive atom. They consist of two neutrons and two protons and are thus identical to the nucleus of a helium atom. Because of their large mass, alpha particles have a short range of only a few centimetres in air, and can be stopped by a sheet of paper. Beta particles are more penetrating and can travel through a 3-mm/0.1-in sheet of aluminium or up to 1 m/3 ft of air. They consist of high-energy electrons emitted at high velocity from a radioactive atom that is undergoing spontaneous disintegration. Gamma rays comprise very high-frequency electromagnetic radiation. Gamma rays are stopped only by direct collision with an atom and are therefore very penetrating; they can, however, be stopped by about 4 cm/1.5 in of lead. When alpha, beta, and gamma radiation pass through matter they tend to knock electrons out of atoms, ionizing them. They are therefore called ionizing radiation. Alpha particles are the most ionizing, being heavy, slow moving and carrying two positive charges. Gamma rays are weakly ionizing as they carry no charge. Beta particles fall between alpha and gamma radiation in ionizing potential.

detection of radioactivity

Detectors of ionizing radiation make use of the ionizing properties of radiation to cause changes that can be detected and measured. A Geiger counter detects the momentary current that passes between electrodes in a suitable gas when ionizing radiation causes the ionization of that gas. The device is named after the German physicist Hans Geiger. The activity of a radioactive source describes the rate at which nuclei are disintegrating within it. One becquerel (1 Bq) is defined as a rate of one disintegration per second.

radioactive decay

Radioactive decay occurs when an unstable nucleus emits alpha, beta, or gamma radiation in order to become more stable. The energy given out by disintegrating atoms is called atomic radiation. An alpha particle consists of two protons and two neutrons. When alpha decay occurs (the emission of an alpha particle from a nucleus) it results in the formation of a new nucleus. An atom of uranium isotope of mass 238, on emitting an alpha particle, becomes an atom of thorium, mass 234. Beta decay, the loss of an electron from an atom, is accomplished by the transformation of a neutron into a proton, thus resulting in an increase in the atomic number of one. For example, the decay of the carbon 314 isotope results in the formation of an atom of nitrogen (mass 14, atomic number 7) and the emission of an electron. Gamma emission usually occurs as part of alpha or beta emission. High-speed electromagnetic radiation is emitted from the nucleus in order to make it more stable during the loss of an alpha or beta particle. Isotopes of an element have different atomic masses. They have the same number of protons but different numbers of neutrons in the nucleus. For example, uranium 235 and uranium 238 both have 92 protons but the latter has three more neutrons than the former. Some isotopes are naturally radioactive (see radioisotopes) while others are not. Radioactive decay can take place either as a one-step decay, or through a series of steps that transmute one element into another. This is called a decay series or chain, and sometimes produces an element more radioactive than its predecessor. For example, uranium 238 decays by alpha emission to thorium 234; thorium 234 is a beta emitter and decays to give protactinium 234. This emits a beta particle to form uranium 234, which in turn undergoes alpha decay to form thorium 230. A further alpha decay yields the isotope radium 226.

health hazards

We are surrounded by radioactive substances. Our food contains traces of radioactive isotopes and our own bodies are made of naturally radioactive matter. In addition, we are bombarded by streams of high-energy charged particles from outer space. Radiation present in the environment is known as background radiation and we should take this into account when considering the risk of exposure to other sources. Alpha, beta, and gamma radiation are dangerous to body tissues because of their ionizing properties, especially if a radioactive substance is ingested or inhaled. Illness resulting from exposure to radioactive substances can take various forms, which are collectively known as radiation sickness.

radioactivity in use

Radioactivity has a number of uses in modern science, but its use should always be carefully controlled and monitored to minimize the risk of harm to living things. In science, a small quantity of a radioactive tracer can be used to follow the path of a chemical reaction or a physical or biological process. Radiocarbon dating is a technique for measuring the age of organic materials. Another application is in determining the age of rocks. This is based on the fact that in many uranium and thorium ores, all of which have been decaying since the formation of the rock, the alpha particles released during decay have been trapped as helium atoms in the rock. The age of the rock can be assessed by calculating the relative amounts of helium, uranium, and thorium in it. This calculation can help to estimate the age of the Earth at around 4.6 billion years. In medicine, radioactive emissions and electromagnetic radiation can be used therapeutically; for example, to treat cancer, when the radiation dose is very carefully controlled.

nuclear fission and fusion

Fission of a nucleus occurs when the nucleus splits into two approximately equal fragments. The fission of the nucleus results in the release of neutrons and a large amount of energy. In a nuclear reactor, the fission of uranium 235 is caused by bombarding it with neutrons. A nuclear chain reaction is caused as neutrons released by the splitting of atomic nuclei themselves go on to split other nuclei, releasing even more neutrons. In a nuclear reactor this process is carefully controlled to release nuclear energy. In nuclear fusion, two light nuclei combine to form a bigger nucleus. As fusion is accompanied by the release of large amounts of energy, the process might one day be harnessed to form the basis of commercial energy production. So far, no successful fusion reactor has been built.

radio frequencies and wavelengths

see electromagnetic waves.

radioisotope or radioactive isotope

naturally occurring or synthetic radioactive form of an element. Most radioisotopes are made by bombarding a stable element with neutrons in the core of a nuclear reactor (see fission). The radiations given off by radioisotopes are easy to detect (hence their use as tracers), can in some instances penetrate substantial thicknesses of materials, and have profound effects (such as genetic mutation) on living matter.

Most natural isotopes of relative atomic mass below 208 are not radioactive. Those from 210 and up are all radioactive.

uses

Radioisotopes have many uses in medicine, for example in radiotherapy and radioisotope scanning. The use of radioactive isotopes in the diagnosis, investigation, and treatment of disease is called *nuclear medicine*.

radio wave

electromagnetic wave possessing a long wavelength (ranging from about 10^{-3} to 10^4 m) and a low frequency (from about 10^5 to 10^{11} Hz). Included in the radio wave part of the spectrum are microwaves, used for both communications and for cooking; ultra high- and very high-frequency waves, used for television and FM (frequency modulation) radio communications; and short, medium, and long waves, used for AM (amplitude modulation) radio communications. Radio waves that are used for communications have all been modulated (see modulation) to carry information. Certain astronomical objects emit radio waves, which may be detected and studied using radio telescopes.

reactance

property of an alternating current circuit that together with any resistance makes up the impedance (the total opposition of the circuit to the passage of a current).

The reactance to an alternating current of frequency f of an inductance L is $2\pi f L$ and that of a capacitance C is $1/2\pi f C$. Reactance is measured in ohms.

rectifier
device used for obtaining one-directional current (DC) from an alternating source of supply (AC). (The process is necessary because almost all electrical power is generated, transmitted, and supplied as alternating current, but many devices, from television sets to electric motors, require direct current.) Types include plate rectifiers, thermionic diodes, and semiconductor diodes.

A single diode produces half-wave rectification in which current flows in one direction for one-half of the alternating-current cycle only – an inefficient process in which the power is effectively turned off for half the time. A *bridge rectifier*, constructed from four diodes, can rectify the alternating supply in such a way that both the positive and negative halves of the alternating cycle can produce a current flowing in the same direction.

reflection
throwing back or deflection of waves, such as light or sound waves, when they hit a surface. The *law of reflection* states that the angle of incidence (the angle between the ray and a perpendicular line drawn to the surface) is equal to the angle of reflection (the angle between the reflected ray and a perpendicular to the surface).

When light passes from a dense medium to a less dense medium, such as from water to air, both refraction and reflection can occur. If the angle of incidence is small, the reflection will be relatively weak compared to the refraction. But as the angle of incidence increases the relative degree of reflection will increase. At some *critical angle of incidence* the angle of refraction is 90°. Since refraction cannot occur above 90°, the light is totally reflected at angles above this critical angle of incidence. This condition is known as *total internal reflection*. Total internal reflection is used in fibre optics to transmit data over long distances, without the need of amplification.

refraction
bending of a wave when it passes from one medium into another. It is the effect of the different speeds of wave propagation in two substances that have different densities. The amount of refraction depends on the densities of the media, the angle at which the wave strikes the surface of the second medium, and the amount of bending and change of velocity corresponding to the wave's frequency (dispersion). Refraction occurs with all types of progressive waves – electromagnetic waves, sound waves, and water waves – and differs from reflection, which involves no change in velocity.

refraction of light
The degree of refraction depends in part on the angle at which the light hits the surface of a material. A line perpendicular to that surface is called the *normal*. The angle between the incoming light ray and the normal to the surface is called

the *angle of incidence*. The angle between the refracted ray and the normal is called the *angle of refraction*. The angle of refraction cannot exceed 90°. An example of refraction is light hitting a glass pane. When light in air enters the denser medium, it is bent toward the normal. When light passes out of the glass into the air, which is less dense, it is bent away from the normal. The incident light will be parallel to the emerging light because the two faces of the glass are parallel. However, if the two faces are not parallel, as with a prism, the emerging light will not be parallel to the incident light. The angle between the incident ray and the emerging ray is called the *angle of deviation*. The amount of bending and change in velocity of the refracted wave is due to the amount of dispersion corresponding to the wave's frequency, and the *refractive index* of the material. When light hits the denser material, its frequency remains constant, but its velocity decreases due to the influence of electrons in the denser medium. Constant frequency means that the same number of light waves must pass by in the same amount of time. If the waves are slowing down, wavelength must also decrease to maintain the constant frequency. The waves become more closely spaced, bending toward the normal as if they are being dragged. The refractive index of a material indicates by how much a wave is bent. It is found by dividing the velocity of the wave in the first medium by the velocity of the wave in the second medium. The *absolute refractive index* of a material is the velocity of light in that material relative to the velocity of light in a vacuum. See also apparent depth.

refraction of sound
Sound waves, unlike light, travel faster in denser materials, such as solids and liquids, than they travel in air. When sound waves enter a solid, their velocity and wavelength *increase* and they are bent away from the normal to the surface of the solid.

water waves
Water waves are refracted when their velocity decreases. Water waves slow down as water becomes shallower. A good example of this is a wavefront approaching a shore that is shallower in one place and deeper in another. When the wavefront approaches, the part of the wave in shallower water will slow down and its wavelength will decrease, causing it to lag behind the part of the wave in the deeper water.

refractive index
measure of the refraction of a ray of light as it passes from one transparent medium to another. If the angle of incidence is i and the angle of refraction is r, the ratio of the two refractive indices is given by $n_1/n_2 = \sin i/\sin r$. It is also equal to the speed of light in the first medium divided by the speed of light in the second, and it varies with the wavelength of the light.

relative atomic mass
mass of an atom relative to one-twelfth the mass of an atom of carbon-12. It depends primarily on the number of protons and neutrons in the atom, the

electrons having negligible mass. If more than one isotope of the element is present, the relative atomic mass is calculated by taking an average that takes account of the relative proportions of each isotope, resulting in values that are not whole numbers. The term *atomic weight*, although commonly used, is strictly speaking incorrect.

relative density
density (at 20°C/68°F) of a solid or liquid relative to (divided by) the maximum density of water (at 4°C/39.2°F). The relative density of a gas is its density divided by the density of hydrogen (or sometimes dry air) at the same temperature and pressure.

relative humidity
concentration of water vapour in the air. It is expressed as the ratio of the partial pressure of the water vapour to its saturated vapour pressure at the same temperature. The higher the temperature, the higher the saturated vapour pressure.

relativity
theory of the relative rather than absolute character of mass, time, and space, and their interdependence, as developed by German-born US physicist Albert Einstein in two phases:

special theory of relativity (1905)
Starting with the premises that (1) the laws of nature are the same for all observers in unaccelerated motion, and (2) the speed of light is independent of the motion of its source, Einstein arrived at some rather unexpected consequences. Intuitively familiar concepts, like mass, length, and time, had to be modified. For example, an object moving rapidly past the observer will appear to be both shorter and more massive than when it is at rest (that is, at rest relative to the observer), and a clock moving rapidly past the observer will appear to be running slower than when it is at rest. These predictions of relativity theory seem to be foreign to everyday experience merely because the changes are quite negligible at speeds less than about 1,500 km s^{-1}, and they only become appreciable at speeds approaching the speed of light.

general theory of relativity (1915)
The geometrical properties of space-time were to be conceived as modified locally by the presence of a body with mass. A planet's orbit around the Sun (as observed in three-dimensional space) arises from its natural trajectory in modified space-time. Einstein's general theory accounts for a peculiarity in the behaviour of the motion of the perihelion of the orbit of the planet Mercury that cannot be explained in Newton's theory. The new theory also said that light rays should bend when they pass by a massive object. The predicted bending of starlight was observed during the eclipse of the Sun in 1919. A third corroboration is found in the shift towards the red in the spectra of the Sun and, in particular, of stars of great density – white dwarfs such as the companion of Sirius.

Einstein showed that, for consistency with the above premises (1) and (2), the principles of dynamics as established by Newton needed modification; the most celebrated new result was the equation $E = mc^2$, which expresses an equivalence between mass (m) and energy (E), c being the speed of light in a vacuum. In 'relativistic mechanics', conservation of mass is replaced by the new concept of conservation of 'mass-energy'. General relativity is central to modern astrophysics and cosmology; it predicts, for example, the possibility of black holes. General relativity theory was inspired by the simple idea that it is impossible in a small region to distinguish between acceleration and gravitation effects (as in a lift one feels heavier when the lift accelerates upwards), but the mathematical development of the idea is formidable. Such is not the case for the special theory, which a nonexpert can follow up to $E = mc^2$ and beyond.

A famous relativity problem is the 'twin paradox'. If one of two twins, O', sets off at high speed in a space ship, travels for some years, and then turns around and returns at high speed to rejoin the other twin, O, who stayed at home, he will have aged less than O, because his clocks have been running slower. If his speed is $v = 0.995c$ in each direction, he will have aged only one year for every 10 years that O has aged. This conclusion has been hotly disputed, because it appears to contradict the symmetry that ought to exist between O and O'. But now there *is* no such symmetry: O has remained in an unaccelerated state, whereas O' has suffered three very large changes in velocity, and both will in fact agree that O has aged more than O'. The observed slowing down of the decay of mesons when moving conclusively shows the existence of this effect, called time dilation.

rem acronym for Roentgen equivalent man
unit of radiation dose equivalent.

The rem has now been replaced in the SI system by the sievert (one rem equals 0.01 sievert), but remains in common use.

resistance
that property of a conductor that restricts the flow of electricity through it, associated with the conversion of electrical energy to heat; also the magnitude of this property. Resistance depends on many factors, such as the nature of the material, its temperature, dimensions, and thermal properties; degree of impurity; the nature and state of illumination of the surface; and the frequency and magnitude of the current. The SI unit of resistance is the ohm.

$$\text{resistance} = \frac{\text{voltage}}{\text{current}}$$

The statement that current is proportional to voltage (reistance is constant) at constant temperature is known as Ohm's law. It is approximately true for many materials that are accordingly described as 'ohmic'.

resistivity

measure of the ability of a material to resist the flow of an electric current. It is numerically equal to the resistance of a sample of unit length and unit cross-sectional area, and its unit is the ohm metre (symbol Ωm). A good conductor has a low resistivity (1.7×10^{-8} Ωm for copper); an insulator has a very high resistivity (10^{15} Ωm for polyethane).

resistor

any component in an electrical circuit used to introduce resistance to a current. Resistors are often made from wire-wound coils or pieces of carbon. Rheostats and potentiometers are variable resistors.

When resistors R_1, R_2, R_3,... are connected in a series circuit, the total resistance of the circuit is $R_1 + R_2 + R_3 + ...$. When resistors R_1, R_2, R_3,... are connected in a parallel circuit, the total resistance of the circuit is R, given by

$$\frac{1}{R} = \frac{1}{R_1} + \frac{1}{R_2} + \frac{1}{R_3} + ...$$

resonance

rapid amplification of a vibration when the vibrating object is subject to a force varying at its natural frequency. In a trombone, for example, the length of the air column in the instrument is adjusted until it resonates with the note being sounded. Resonance effects are also produced by many electrical circuits. Tuning a radio, for example, is done by adjusting the natural frequency of the receiver circuit until it coincides with the frequency of the radio waves falling on the aerial.

Resonance has many physical applications. Children use it to increase the size of the movement on a swing, by giving a push at the same point during each swing. Soldiers marching across a bridge in step could cause the bridge to vibrate violently if the frequency of their steps coincided with its natural frequency. Resonance caused of the collapse of the Tacoma Narrows Bridge, USA, in 1940, when the frequency of the wind gusts coincided with the natural frequency of the bridge.

rest mass

mass of a body when its velocity is zero or considerably below that of light. According to the theory of relativity, at very high velocities, there is a relativistic effect that increases the mass of the particle.

reverberation

multiple reflections, or echoes, of sounds inside a building that merge and persist a short time (up to a few seconds) before fading away. At each reflection some of the sound energy is absorbed, causing the amplitude of the sound wave and the intensity of the sound to reduce a little.

Too much reverberation causes sounds to become confused and indistinct, and this is particularly noticeable in empty rooms and halls, and such buildings as churches and cathedrals where the hard, unfurnished surfaces do not

241

absorb sound energy well. Where walls and surfaces absorb sound energy very efficiently, too little reverberation may cause a room or hall to sound dull or 'dead'. Reverberation is a key factor in the design of theatres and concert halls, and can be controlled by lining ceilings and walls with materials possessing specific sound-absorbing properties.

rheostat

variable resistor, usually consisting of a high-resistance wire-wound coil with a sliding contact. It is used to vary electrical resistance without interrupting the current (for example, when dimming lights). The circular type, which can be used, for example, as the volume control of an amplifier, is also known as a potentiometer.

scalar quantity

in mathematics and science, a quantity that has magnitude but no direction, as distinct from a vector quantity, which has a direction as well as a magnitude. Temperature, mass, and volume are scalar quantities.

scattering

random deviation or reflection of a stream of particles or of a beam of radiation such as light, by the particles in the matter through which it passes.

alpha particles

Alpha particles scattered by a thin gold foil provided the first convincing evidence that atoms had very small, very dense, positive nuclei. From 1906 to 1908 Ernest Rutherford carried out a series of experiments from which he estimated that the closest approach of an alpha particle to a gold nucleus in a head-on collision was about 10^{-14} m. He concluded that the gold nucleus must be no larger than this. Most of the alpha particles fired at the gold foil passed straight through undeviated; however, a few were scattered in all directions and a very small fraction bounced back towards the source. This result so surprised Rutherford that he is reported to have commented: 'It was almost as if you fired a 15-inch shell at a piece of tissue paper and it came back and hit you'.

light

Light is scattered from a rough surface, such as that of a sheet of paper, by random reflection from the varying angles of each small part of the surface. This is responsible for the dull, flat appearance of such surfaces and their inability to form images (unlike mirrors). Light is also scattered by particles suspended in a gas or liquid. The red and yellow colours associated with sunrises and sunsets are due to the fact that red light is scattered to a lesser extent than is blue light by dust particles in the atmosphere. When the Sun is low in the sky, its light passes through a thicker, more dusty layer of the atmosphere, and the blue light radiated by it is scattered away, leaving the red sunlight to pass through to the eye of the observer.

secondary emission

in physics, an emission of electrons from the surface of certain substances when they are struck by high-speed electrons or other particles from an external source. It can be detected with a photomultiplier.

Seebeck effect

in physics, the generation of a voltage in a circuit containing two different metals, or semiconductors, by keeping the junctions between them at different temperatures. Discovered by the German physicist Thomas Seebeck (1770–1831), it is also called the thermoelectric effect, and is the basis of the thermocouple. It is the opposite of the Peltier effect (in which current flow causes a temperature difference between the junctions of different metals).

self-inductance or self-induction

in physics, the creation of an electromotive force opposing the current. See inductance.

semiconductor

material with electrical conductivity intermediate between metals and insulators and used in a wide range of electronic devices. Certain crystalline materials, most notably silicon and germanium, have a small number of free electrons that have escaped from the bonds between the atoms. The atoms from which they have escaped possess vacancies, called holes, which are similarly able to move from atom to atom and can be regarded as positive charges. Current can be carried by both electrons (negative carriers) and holes (positive carriers). Such materials are known as *intrinsic semiconductors*.

Conductivity of a semiconductor can be enhanced by doping the material with small numbers of impurity atoms which either release free electrons (making an *n-type semiconductor* with more electrons than holes) or capture them (a *p-type semiconductor* with more holes than electrons). When p-type and n-type materials are brought together to form a p–n junction, an electrical barrier is formed that conducts current more readily in one direction than the other. This is the basis of the semiconductor diode, used for rectification, and numerous other devices including transistors, rectifiers, and integrated circuits (silicon chips).

semiconductor diode or p–n junction diode

two-terminal semiconductor device that allows electric current to flow in only one direction, the *forward-bias* direction. A very high resistance prevents current flow in the opposite, or *reverse-bias*, direction. It is used as a rectifier, converting alternating current (AC) to direct current (DC).

The diode is cut from a single crystal of a semiconductor (such as silicon) to which special impurities have been added during manufacture so that the crystal is now composed of two distinct regions. One region contains semiconductor material of the *p*-type, which contains more *positive* charge carriers than negative; the other contains material of the *n*-type, which has more *n*egative charge carriers than positive. The region of contact between the two types is called the *p–n* junction, and it is this that acts as the barrier preventing current from flowing, in conventional current terms, from the *n*-type to the *p*-type (in the reverse-bias direction).

series circuit

electrical circuit in which the components are connected end to end, so that the current flows through them all one after the other.

shadow

area of darkness behind an opaque object that cannot be reached by some or all of the light coming from a light source in front. Its presence may be explained in terms of light rays travelling in straight lines and being unable to bend round obstacles. A point source of light produces an umbra, a completely black shadow with sharp edges. An extended source of light produces both a central umbra and a penumbra, a region of semidarkness with blurred edges where darkness gives way to light.

short circuit

unintended direct connection between two points in an electrical circuit. Resistance is proportional to the length of wire through which current flows. By bypassing the rest of the circuit, the short circuit has low resistance and a large current flows through it. This may cause the circuit to overheat dangerously.

siemens

SI unit (symbol S) of electrical conductance, the reciprocal of the resistance of an electrical circuit. One siemens equals one ampere per volt. It was formerly called the mho or reciprocal ohm.

sievert

SI unit (symbol Sv) of radiation dose equivalent. It replaces the rem (1 Sv equals 100 rem). Some types of radiation do more damage than others for the same absorbed dose – for example, an absorbed dose of alpha radiation causes 20 times as much biological damage as the same dose of beta radiation. The equivalent dose in sieverts is equal to the absorbed dose of radiation in grays multiplied by the relative biological effectiveness. Humans can absorb up to 0.25 Sv without immediate ill effects; 1 Sv may produce radiation sickness; and more than 8 Sv causes death.

smart fluid or electrorheological fluid

liquid suspension that solidifies to form a jellylike solid when a high-voltage electric field is applied across it and that returns to the liquid state when the field is removed. Most smart fluids are zeolites or metals coated with polymers or oxides.

Snell's law of refraction

in optics, the rule that when a ray of light passes from one medium to another, the sine of the angle of incidence divided by the sine of the angle of refraction is equal to the ratio of the indices of refraction in the two media. For a ray passing from medium 1 to medium 2: $n_2/n_1 = \sin i/\sin r$ where n_1 and n_2 are the refractive indices of the two media. The law was devised by the Dutch physicist, Willebrord Snell.

solar radiation

radiation given off by the Sun, consisting mainly of visible light, ultraviolet radiation, and infrared radiation, although the whole spectrum of electromagnetic waves is present, from radio waves to X-rays. High-energy charged particles, such as electrons, are also emitted, especially from solar flares. When these reach the Earth, they cause magnetic storms (disruptions of the Earth's magnetic field), which interfere with radio communications.

solenoid

coil of wire, usually cylindrical, in which a magnetic field is created by passing an electric current through it (see electromagnet). This field can be used to move an iron rod placed on its axis. Mechanical valves attached to the rod can be operated by switching the current on or off, so converting electrical energy into mechanical energy. Solenoids are used to relay energy from the battery of a car to the starter motor by means of the ignition switch.

solid

state of matter that holds its own shape (as opposed to a liquid, which takes up the shape of its container, or a gas, which totally fills its container). According to kinetic theory, the atoms or molecules in a solid are not free to move but merely vibrate about fixed positions, such as those in crystal lattices.

sonar acronym for sound navigation and ranging

method of locating underwater objects by the reflection of ultrasonic waves. The time taken for an acoustic beam to travel to the object and back to the source enables the distance to be found since the velocity of sound in water is known. Sonar devices, or *echo sounders*, were developed in 1920, and are the commonest means of underwater navigation.

sound

physiological sensation received by the ear, originating in a vibration that communicates itself as a pressure variation in the air and travels in every direction, spreading out as an expanding sphere. All sound waves in air travel with a speed dependent on the temperature; under ordinary conditions, this is about 330 m/1,070 ft per second. The pitch of the sound depends on the number of vibrations imposed on the air per second (frequency), but the speed is unaffected. The loudness of a sound is dependent primarily on the amplitude of the vibration of the air.

Sound travels as a *longitudinal wave*, that is, its compressions and rarefactions are in the direction of propagation. Like other waves – light waves and water waves – sound can be reflected, diffracted, and refracted. Reflection of a sound wave is heard as an *echo*. Diffraction explains why sound can be heard round doorways. When sound is refracted (see refraction), sound is bent when it passes into a denser or less dense material because sound travels faster in denser materials, such as solids and liquids. The lowest note audible to a human being has a frequency of about 20 hertz (vibrations per second), and the highest one of about 20,000 Hz; the lower limit of this range varies little with the person's age, but the upper range falls steadily from adolescence onwards.

space-time
combination of space and time used in the theory of relativity. When developing relativity, Albert Einstein showed that time was in many respects like an extra dimension (or direction) to space. Space and time can thus be considered as entwined into a single entity, rather than two separate things.

Space-time is considered to have four dimensions: three of space and one of time. In relativity theory, events are described as occurring at points in space-time. The *general theory of relativity* describes how space-time is distorted by the presence of material bodies, an effect that we observe as gravity.

spark chamber
electronic device for recording tracks of charged subatomic particles, decay products, and rays. In combination with a stack of photographic plates, a spark chamber enables the point where an interaction has taken place to be located, to within a cubic centimetre. At its simplest, it consists of two smooth thread-like electrodes that are positioned 1–2 cm/0.5–1 in apart, the space between being filled by an inert gas such as neon. Sparks jump through the gas along the ionized path created by the radiation. See particle detector.

specific gravity
alternative term for *relative density*.

specific heat capacity
quantity of heat required to raise unit mass (1 kg) of a substance by one kelvin (1 K). The unit of specific heat capacity in the SI system is the joule per kilogram kelvin ($J kg^{-1} K^{-1}$).

spectrometer
instrument used to study the composition of light emitted by a source. The range, or spectrum, of wavelengths emitted by a source depends upon its constituent elements, and may be used to determine its chemical composition.

The simpler forms of spectrometer analyse only visible light. A *collimator* receives the incoming rays and produces a parallel beam, which is then split into a spectrum by either a diffraction grating or a prism mounted on a turntable. As the turntable is rotated each of the constituent colours of the beam may be seen through a *telescope*, and the angle at which each has been deviated may be measured on a circular scale. From this information the wavelengths of the colours of light can be calculated.

spectroscopy

study of spectra (see spectrum) associated with atoms or molecules in solid, liquid, or gaseous phase. Spectroscopy can be used to identify unknown compounds and is an invaluable tool in science, medicine, and industry (for example, in checking the purity of drugs).

Emission spectroscopy is the study of the characteristic series of sharp lines in the spectrum produced when an element is heated. Thus an unknown mixture can be analysed for its component elements. Related is *absorption spectroscopy*, dealing with atoms and molecules as they absorb energy in a characteristic way. Again, dark lines can be used for analysis. More detailed structural information can be obtained using *infrared spectroscopy* (concerned with molecular vibrations) or *nuclear magnetic resonance (NMR) spectroscopy* (concerned with interactions between adjacent atomic nuclei). *Supersonic jet laser beam spectroscopy* enables the isolation and study of clusters in the gas phase. A laser vaporizes a small sample, which is cooled in helium, and ejected into an evacuated chamber. The jet of clusters expands supersonically, cooling the clusters to near absolute zero, and stabilizing them for study in a mass spectrometer.

spectrum plural spectra

pattern of frequencies or wavelengths obtained when electromagnetic radiations are separated into their constituent parts. Visible light is part of the electromagnetic spectrum and most sources emit waves over a range of wavelengths that can be broken up or 'dispersed'; white light can be separated into red, orange, yellow, green, blue, indigo, and violet. The visible spectrum was first studied by Isaac Newton, who showed in 1672 how white light could be broken up into different colours.

There are many types of spectra, both emission and absorption, for radiation and particles, used in spectroscopy. An incandescent body gives rise to a *continuous spectrum* where the dispersed radiation is distributed uninterruptedly over a range of wavelengths. A gaseous element gives a *line spectrum* – one or more bright discrete lines at characteristic wavelengths. Molecular gases give *band spectra* in which there are groups of close-packed lines. In an *absorption spectrum* dark lines or spaces replace the characteristic bright lines of the absorbing medium. The *mass spectrum* of an element is obtained from a mass spectrometer and shows the relative proportions of its constituent isotopes.

speed

rate at which an object moves. The average speed v of an object may be calculated by dividing the distance s it has travelled by the time t taken to do so, and may be expressed as:

$$v = \frac{s}{t}$$

The usual units of speed are metres per second or kilometres per hour.

Speed is a scalar quantity in which direction of motion is unimportant (unlike the vector quantity velocity, in which both magnitude and direction must be taken into consideration). See also distance-time graph and speed-time graph.

speed of light
speed at which light and other electromagnetic waves travel through empty space. Its value is 299,792,458 m/186,282 mi per second; for most calculations 3×10^8 m per second suffices. The speed of light is the highest speed possible, according to the theory of relativity, and its value is independent of the motion of its source and of the observer. It is impossible to accelerate any material body to this speed because it would require an infinite amount of energy.

speed of sound
speed at which sound travels through a medium, such as air or water. In air at a temperature of 0°C/32°F, the speed of sound is 331 m/1,087 ft per second. At higher temperatures, the speed of sound is greater; at 18°C/64°F it is 342 m/1,123 ft per second. It is greater in liquids and solids; for example, in water it is around 1,440 m/4,724 ft per second, depending on the temperature.

speed–time graph
graph used to describe the motion of a body by illustrating how its speed or velocity changes with time. The gradient of the graph gives the object's acceleration: if the gradient is zero (the graph is horizontal) then the body is moving with constant speed or uniform velocity; if the gradient is constant, the body is moving with uniform acceleration. The area under the graph gives the total distance travelled by the body.

spin
intrinsic angular momentum of a subatomic particle, nucleus, atom, or molecule, which continues to exist even when the particle comes to rest. A particle in a specific energy state has a particular spin, just as it has a particular electric charge and mass. According to quantum theory, this is restricted to discrete and indivisible values, specified by a spin quantum number. Because of its spin, a charged particle acts as a small magnet and is affected by magnetic fields.

standard form or scientific notation
method of writing numbers often used by scientists, particularly for very large or very small numbers. The numbers are written with one digit before the decimal point and multiplied by a power of 10. The number of digits given after the decimal point depends on the accuracy required. For example, the speed of light is 2.9979×10^8 m/1.8628×10^5 mi per second.

standard model
modern theory of elementary particles and their interactions. According to the standard model, elementary particles are classified as leptons (light particles, such as electrons), hadrons (particles, such as neutrons and protons, that are

formed from quarks), and gauge bosons. Leptons and hadrons interact by exchanging gauge bosons, each of which is responsible for a different fundamental force: photons mediate the electromagnetic force, which affects all charged particles; gluons mediate the strong nuclear force, which affects quarks; gravitons mediate the force of gravity; and the intermediate vector bosons mediate the weak nuclear force. See also forces, fundamental, quantum electrodynamics, and quantum chromodynamics.

standing wave

wave in which the positions of nodes (positions of zero vibration) and antinodes (positions of maximum vibration) do not move. Standing waves result when two similar waves travel in opposite directions through the same space.

For example, when a sound wave is reflected back along its own path, as when a stretched string is plucked, a standing wave is formed. In this case the antinode remains fixed at the centre and the nodes are at the two ends. Water and electromagnetic waves can form standing waves in the same way.

states of matter

forms (solid, liquid, or gas) in which material can exist. Whether a material is solid, liquid, or gaseous depends on its temperature and the pressure on it. The transition between states takes place at definite temperatures, called melting point and boiling point.

Kinetic theory describes how the state of a material depends on the movement and arrangement of its atoms or molecules. The atoms or molecules of *gases* move randomly in otherwise empty space, filling any size or shape of container. Gases can be liquefied by cooling as this lowers the speed of the molecules and enables attractive forces between them to bind them together. A *liquid* forms a level surface and assumes the shape of its container; its atoms or molecules do not occupy fixed positions, nor do they have total freedom of movement. *Solids* hold their own shape as their atoms or molecules are not free to move about but merely vibrate about fixed positions, such as those in crystal lattices.

A hot ionized gas or plasma is often called the fourth state of matter, but liquid crystals, colloids, and glass also have a claim to this title.

change of state

This occurs when a gas condenses to a liquid or a liquid freezes to a solid. Similar changes take place when a solid melts to form a liquid or a liquid vaporizes (evaporates) to form a gas. The first set of changes are brought about by cooling, the second set by heating. In the unusual change of state called *sublimation*, a solid changes directly to a gas without passing through the liquid state. For example, solid carbon dioxide (dry ice) sublimes to carbon dioxide gas.

static electricity

electric charge that is stationary, usually acquired by a body by means of electrostatic induction or friction. Rubbing different materials can produce static electricity, as seen in the sparks produced on combing one's hair or removing a nylon shirt. In some processes static electricity is useful, as in paint spraying where the parts to be sprayed are charged with electricity of opposite polarity to that on the paint droplets, and in xerography.

statics
branch of mechanics concerned with the behaviour of bodies at rest and forces in equilibrium, and distinguished from dynamics.

statistical mechanics
branch of physics in which the properties of large collections of particles are predicted by considering the motions of the constituent particles. It is closely related to thermodynamics.

Stefan–Boltzmann constant
constant relating the energy emitted by a black body (a hypothetical body that absorbs or emits all the energy falling on it) to its temperature. Its value is 5.6697×10^{-8} W m^{-2} K^{-4}.

stress and strain
measures of the deforming force applied to a body (stress) and of the resulting change in its shape (strain). For a perfectly elastic material, stress is proportional to strain (Hooke's law).

string theory
mathematical theory developed in the 1980s to explain the behaviour of elementary particles; see superstring theory.

stroboscope
instrument for studying continuous periodic motion by using light flashing at the same frequency as that of the motion; for example, rotating machinery can be optically 'stopped' by illuminating it with a stroboscope flashing at the exact rate of rotation.

strong nuclear force
one of the four fundamental forces of nature, the other three being the gravitational force or gravity, the electromagnetic force, and the weak nuclear force. The strong nuclear force was first described by the Japanese physicist Hideki Yukawa in 1935. It is the strongest of all the forces, acts only over very small distances within the nucleus of the atom (10^{-13} cm), and is responsible for binding together quarks to form hadrons, and for binding together protons and neutrons in the atomic nucleus. The particle that is the carrier of the strong nuclear force is the gluon, of which there are eight kinds, each with zero mass and zero charge.

subatomic particle
particle that is smaller than an atom. Such particles may be indivisible elementary particles, such as the electron and quark, or they may be composites, such as the proton, neutron, and alpha particle. See also particle physics.

superconductivity

increase in electrical conductivity at low temperatures. The resistance of some metals and metallic compounds decreases uniformly with decreasing temperature until at a critical temperature (the superconducting point), within a few degrees of absolute zero (0 K/–273.15°C/–459.67°F), the resistance suddenly falls to zero. The phenomenon was discovered by Dutch scientist Heike Kamerlingh Onnes in 1911.

Some metals, such as platinum and copper, do not become superconductive; as the temperature decreases, their resistance decreases to a certain point but then rises again. Superconductivity can be nullified by the application of a large magnetic field. In the superconducting state, an electric current will continue indefinitely once started, provided that the material remains below the superconducting point. In 1986 IBM researchers achieved superconductivity with some ceramics at –243°C/–405°F, opening up the possibility of *'high-temperature' superconductivity*, Paul Chu at the University of Houston, Texas, USA, achieved superconductivity at –179°C/–290°F, a temperature that can be sustained using liquid nitrogen. Researchers are now trying to find a material that will be superconducting at room temperature.

supercooling

cooling of a liquid below its freezing point without freezing taking place; or the cooling of a saturated solution without crystallization taking place, to form a supersaturated solution. In both cases supercooling is possible because of the lack of solid particles around which crystals can form. Crystallization rapidly follows the introduction of a small crystal (seed) or agitation of the supercooled solution.

supercritical fluid

fluid that combines the properties of a gas and a liquid, see fluid, supercritical.

superfluid

fluid that flows without viscosity or friction and has a very high thermal conductivity. Liquid helium at temperatures below 2 K (–271°C/–456°F) is a superfluid: it shows unexpected behaviour; for instance, it flows uphill in apparent defiance of gravity and, if placed in a container, will flow up the sides and escape.

German physicists discovered in 1998 that as few as 60 atoms will exhibit superfluidity.

supersonic speed

speed greater than that at which sound travels, measured in Mach numbers. In dry air at 0°C/32°F, sound travels at about 1,170 kph/727 mph, but decreases its speed with altitude until, at 12,000 m/39,000 ft, it is only 1,060 kph/658 mph.

When an aircraft passes the sound barrier, shock waves are built up that give rise to sonic boom, often heard at ground level. US pilot Captain Charles Yeager was the first to achieve supersonic flight, in a Bell VS-1 rocket plane on 14 October 1947.

superstring theory
in physics, a mathematical theory developed in the 1980s to explain the properties of elementary particles and the forces between them (in particular, gravity and the nuclear forces) in a way that combines relativity and quantum theory. In string theory, the fundamental objects in the universe are not pointlike particles but extremely small stringlike objects. These objects exist in a universe of ten dimensions, but since the earliest moments of the Big Bang six of these have been compacted or 'rolled up', so that now, only three space dimensions and one dimension of time are discernible.

There are many unresolved difficulties with superstring theory, but some physicists think it may be the ultimate 'theory of everything' that explains all aspects of the universe within one framework.

supersymmetry
theory that relates the two classes of elementary particle, the fermions and the bosons. According to supersymmetry, each fermion particle has a boson partner particle, and vice versa. It has not been possible to marry up all the known fermions with the known bosons, and so the theory postulates the existence of other, as yet undiscovered fermions, such as the photinos (partners of the photons), gluinos (partners of the gluons), and gravitinos (partners of the gravitons). Using these ideas, it has become possible to develop a theory of gravity – called *supergravity* – that extends Einstein's work and considers the gravitational, nuclear, and electromagnetic forces to be manifestations of an underlying superforce. Supersymmetry has been incorporated into the superstring theory, and appears to be a crucial ingredient in the 'theory of everything' sought by scientists.

surface tension
property that causes the surface of a liquid to behave as if it were covered with a weak elastic skin; this is why a needle can float on water. It is caused by the exposed surface's tendency to contract to the smallest possible area because of cohesive forces between molecules at the surface. Allied phenomena include the formation of droplets, the concave profile of a meniscus, and the capillary action by which water soaks into a sponge.

synchrotron
particle accelerator in which particles move, at increasing speed, around a hollow ring. The particles are guided around the ring by electromagnets, and accelerated by electric fields at points around the ring. Synchrotrons come in a wide range of sizes, the smallest being about 1 m/3.3 ft across while the largest is 27 km/17 mi across. The Tevatron synchrotron at Fermilab is some 6 km/4 mi in circumference and accelerates protons and antiprotons to 1 TeV.

The European Synchrotron Radiation Facility (ESRF) opened in Grenoble, France, in September 1994, funded by £400 million from 12 European countries.

tau
elementary particle with the same electric charge as the electron but a mass nearly double that of a proton. It has a lifetime of around 3×10^{-13} seconds

and belongs to the lepton family of particles – those which interact via the electromagnetic, weak nuclear, and gravitational forces, but not the strong nuclear force.

temperature
degree or intensity of heat of an object and the condition that determines whether it will transfer heat to another object or receive heat from it, according to the laws of thermodynamics. The temperature of an object is a measure of the average kinetic energy possessed by the atoms or molecules of which it is composed. The SI unit of temperature is the kelvin (symbol K) used with the Kelvin scale. Other measures of temperature in common use are the Celsius scale and the Fahrenheit scale.

terminal voltage
potential difference (pd) or voltage across the terminals of a power supply, such as a battery of cells. When the supply is not connected in circuit its terminal voltage is the same as its electromotive force (emf); however, as soon as it begins to supply current to a circuit its terminal voltage falls because some electric potential energy is lost in driving current against the supply's own internal resistance. As the current flowing in the circuit is increased the terminal voltage of the supply falls.

theory
set of ideas, concepts, principles, or methods used to explain a wide set of observed facts. Among the major theories of science are relativity, quantum theory, evolution, and plate tectonics.

Theory of Everything (ToE)
another name for *grand unified theory.*

thermal capacity
another name for *heat capacity*

thermal conductivity
ability of a substance to conduct heat. Good thermal conductors, like good electrical conductors, are generally materials with many free electrons (such as metals).

Thermal conductivity is expressed in units of joules per second per metre per kelvin ($J s^{-1} m^{-1} K^{-1}$). For a block of material of cross-sectional area a and length l, with temperatures T_1 and T_2 at its end faces, the thermal conductivity λ equals $Hl/at(T_2 - T_1)$, where H is the amount of heat transferred in time t.

thermal expansion
expansion that is due to a rise in temperature. It can be expressed in terms of linear, area, or volume expansion.

The coefficient of linear expansion α is the fractional increase in length per degree temperature rise; area, or superficial, expansion β is the fractional increase in area per degree; and volume, or cubic, expansion γ is the fractional increase in volume per degree. To a close approximation, $\beta = 2\alpha$ and $\gamma = 3\alpha$.

thermal reactor
nuclear reactor in which the neutrons released by fission of uranium-235 nuclei are slowed down in order to increase their chances of being captured by other uranium-235 nuclei, and so induce further fission. The material (commonly graphite or heavy water) responsible for doing so is called a ***moderator***. When the fast newly-emitted neutrons collide with the nuclei of the moderator's atoms, some of their kinetic energy is lost and their speed is reduced. Those that have been slowed down to a speed that matches the thermal (heat) energy of the surrounding material are called ***thermal neutrons***, and it is these that are most likely to induce fission and ensure the continuation of the chain reaction. See nuclear reactor and nuclear energy.

thermionics
branch of electronics dealing with the emission of electrons from matter under the influence of heat.

The ***thermionic valve*** (electron tube), used in radio and radar, is a device using space conduction by thermionically emitted electrons from an electrically heated cathode. In most applications valves have been replaced by transistors.

thermodynamics
branch of physics dealing with the transformation of heat into and from other forms of energy. It is the basis of the study of the efficient working of engines, such as the steam and internal-combustion engines. The three laws of thermodynamics are: (1) energy can be neither created nor destroyed, heat and mechanical work being mutually convertible; (2) it is impossible for an unaided self-acting machine to convey heat from one body to another at a higher temperature; and (3) it is impossible by any procedure, no matter how idealized, to reduce any system to the absolute zero of temperature (0 K/−273.15°C/−459.67°F) in a finite number of operations. Put into mathematical form, these laws have widespread applications in physics and chemistry.

thermoluminescence (TL)
release, in the form of a light pulse, of stored nuclear energy in a mineral substance when heated to perhaps 500°C. The energy originates from the radioactive decay of uranium and thorium, and is absorbed by crystalline inclusions within the mineral matrix, such as quartz and feldspar. The release of TL from these crystalline substances is used in archaeology to date pottery, and by geologists in studying terrestrial rocks and meteorites.

thermopile
instrument for measuring radiant heat, consisting of a number of thermocouples connected in series with alternate junctions exposed to the radiation. The current generated (measured by an ammeter) is proportional to the radiation falling on the device.

thyristor
type of rectifier, an electronic device that conducts electricity in one direction only. The thyristor is composed of layers of semiconductor material sandwiched between two electrodes called the anode and cathode. The current can be switched on by using a third electrode called the gate.

Thyristors are used to control mains-driven motors and in lighting dimmer controls.

torque
turning effect of force on an object. A turbine produces a torque that turns an electricity generator in a power station. Torque is measured by multiplying the force by its perpendicular distance from the turning point.

torsion
state of strain set up in a twisted material; for example, when a thread, wire, or rod is twisted, the torsion set up in the material tends to return the material to its original state. The *torsion balance*, a sensitive device for measuring small gravitational or magnetic forces, or electric charges, balances these against the restoring force set up by them in a torsion suspension.

transducer
device that converts one form of energy into another. For example, a thermistor is a transducer that converts heat into an electrical voltage, and an electric motor is a transducer that converts an electrical voltage into mechanical energy. Transducers are important components in many types of sensor, converting the physical quantity to be measured into a proportional voltage signal.

transformer
device in which, by electromagnetic induction, an alternating current (AC) of one voltage is transformed to another voltage, without change of frequency. Transformers are widely used in electrical apparatus of all kinds, and in particular in power transmission where high voltages and low currents are utilized.

A transformer has two coils, a primary for the input and a secondary for the output, wound on a common iron core. The ratio of the primary to the secondary voltages is directly proportional to the number of turns in the primary and secondary coils; the ratio of the current, is inversely proportional.

transistor
solid-state electronic component, made of semiconductor material, with three or more electrodes, that can regulate a current passing through it. A

transistor can act as an amplifier, oscillator, photocell, or switch, and (unlike earlier thermionic valves) usually operates on a very small amount of power. Transistors commonly consist of a tiny sandwich of germanium or silicon, alternate layers having different electrical properties because they are impregnated with minute amounts of different impurities.

A crystal of pure germanium or silicon would act as an insulator (nonconductor). By introducing impurities in the form of atoms of other materials (for example, boron, arsenic, or indium) in minute amounts, the layers may be made either *n-type*, having an excess of electrons, or *p-type*, having a deficiency of electrons. This enables electrons to flow from one layer to another in one direction only. Transistors have had a great impact on the electronics industry, and thousands of millions are now made each year. They perform many of the functions of the thermionic valve, but have the advantages of greater reliability, long life, compactness, and instantaneous action, no warming-up period being necessary. They are widely used in most electronic equipment, including portable radios and televisions, computers, and satellites, and are the basis of the integrated circuit (silicon chip). They were invented at Bell Telephone Laboratories in the USA in 1948 by John Bardeen and Walter Brattain, developing the work of William Shockley.

transverse wave
wave in which the displacement of the medium's particles, or in electromagnetic waves, the direction of the electric and magnetic fields, is at right angles to the direction of travel of the wave motion.

ultrasonics
branch of physics dealing with the theory and application of ultrasound: sound waves occurring at frequencies too high to be heard by the human ear (that is, above about 20 kHz).

The earliest practical application of ultrasonics was the detection of submarines during World War I by reflecting pulses of sound from them (see sonar). Similar principles are now used in industry for nondestructive testing of materials and in medicine to produce images of internal organs and developing fetuses (ultrasound scanning). High-power ultrasound can be used for cleaning, welding plastics, and destroying kidney stones without surgery.

ultraviolet radiation
electromagnetic radiation invisible to the human eye, of wavelengths from about 400 to 4 nm (where the X-ray range begins). Physiologically, ultraviolet radiation is extremely powerful, producing sunburn and causing the formation of vitamin D in the skin.

Ultraviolet rays are strongly germicidal and may be produced artificially by mercury vapour and arc lamps for therapeutic use. The radiation may be detected with ordinary photographic plates or films. It can also be studied by its fluorescent effect on certain materials. The desert iguana *Disposaurus dorsalis* uses it to locate the boundaries of its territory and to find food.

uncertainty principle or indeterminacy principle

in quantum mechanics, the principle that it is impossible to know with unlimited accuracy the position and momentum of a particle. The principle arises because in order to locate a particle exactly, an observer must bounce light (in the form of a photon) off the particle, which must alter its position in an unpredictable way.

It was established by German physicist Werner Heisenberg, and gave a theoretical limit to the precision with which a particle's momentum and position can be measured simultaneously: the more accurately the one is determined, the more uncertainty there is in the other.

unified field theory

theory that attempts to explain the four fundamental forces (strong nuclear, weak nuclear, electromagnetic, and gravity) in terms of a single unified force (see particle physics).

Research was begun by Albert Einstein, and by 1971 a theory developed by US physicists Steven Weinberg and Sheldon Glashow, Pakistani physicist Abdus Salam, and others, had demonstrated the link between the weak and electromagnetic forces. The next stage is to develop a theory (called the grand unified theory) that combines the strong nuclear force with the electroweak force. The final stage will be to incorporate gravity into the scheme. Work on the superstring theory indicates that this may be the ultimate 'theory of everything'.

UV

abbreviation for *ultraviolet.*

vacuum

region completely empty of matter; in physics, any enclosure in which the gas pressure is considerably less than atmospheric pressure (101,325 pascals).

van de Graaff generator

electrostatic generator capable of producing a voltage of over a million volts. It consists of a continuous vertical conveyor belt that carries electrostatic charges (resulting from friction) up to a large hollow sphere supported on an insulated stand. The lower end of the belt is earthed, so that charge accumulates on the sphere. The size of the voltage built up in air depends on the radius of the sphere, but can be increased by enclosing the generator in an inert atmosphere, such as nitrogen.

vapour density

density of a gas, expressed as the mass of a given volume of the gas divided by the mass of an equal volume of a reference gas (such as hydrogen or air) at the same temperature and pressure. If the reference gas is hydrogen, it is equal to half the relative molecular weight (mass) of the gas.

vapour pressure

pressure of a vapour given off by (evaporated from) a liquid or solid, caused by atoms or molecules continuously escaping from its surface. In an enclosed

A modern van de Graaff electrical generator under construction at the Harwell nuclear research site, part of the British Atomic Energy Authority. AEA Technology

space, a maximum value is reached when the number of particles leaving the surface is in equilibrium with those returning to it; this is known as the *saturated vapour pressure* or *equilibrium vapour pressure*.

velocity
speed of an object in a given direction. Velocity is a vector quantity, since its direction is important as well as its magnitude (or speed).

The velocity at any instant of a particle travelling in a curved path is in the direction of the tangent to the path at the instant considered. The velocity v of an object travelling in a fixed direction may be calculated by dividing the distance s it has travelled by the time t taken to do so, and may be expressed as:

$$v = \frac{s}{t}$$

viscosity
resistance of a fluid to flow, caused by its internal friction, which makes it resist flowing past a solid surface or other layers of the fluid. It applies to the motion of an object moving through a fluid as well as the motion of a fluid passing by an object.

Fluids such as pitch, treacle, and heavy oils are highly viscous; for the purposes of calculation, many fluids in physics are considered to be perfect, or nonviscous.

volt
SI unit of electromotive force or electric potential (see potential, electric), symbol V. A small battery has a potential of 1.5 volts, whilst a high-tension transmission line may carry up to 765,000 volts. The domestic electricity supply in the UK is 230 volts (lowered from 240 volts in 1995); it is 110 volts in the USA.

The absolute volt is defined as the potential difference necessary to produce a current of one ampere through an electric circuit with a resistance of one ohm. It can also be defined as the potential difference that requires one joule of work to move a positive charge of one coulomb from the lower to the higher potential. It is named after the Italian scientist Alessandro Volta.

voltage
commonly used term for potential difference (pd) or electromotive force (emf).

voltage amplifier
electronic device that increases an input signal in the form of a voltage or potential difference, delivering an output signal that is larger than the input by a specified ratio.

watt
SI unit (symbol W) of power (the rate of expenditure or consumption of energy) defined as one joule per second. A light bulb, for example, may use 40, 60, 100, or 150 watts of power; an electric heater will use several kilowatts (thousands of watts). The watt is named after the Scottish engineer James Watt.

The absolute watt is defined as the power used when one joule of work is done in one second. In electrical terms, the flow of one ampere of current through a conductor whose ends are at a potential difference of one volt uses one watt of power (watts = volts × amperes).

wave
oscillation that is propagated from a source. Mechanical waves require a medium through which to travel. Electromagnetic waves do not; they can travel

through a vacuum. Waves carry energy but they do not transfer matter. There are two types: in a longitudinal wave, such as a sound wave, the disturbance is parallel to the wave's direction of travel; in a transverse wave, such as an electromagnetic wave, it is perpendicular. The medium (for example the Earth, for seismic waves) is not permanently displaced by the passage of a wave. See also standing wave.

types of wave
There are various ways of classifying wave types. One of these is based on the way the wave travels. In a transverse wave, the displacement of the medium is perpendicular to the direction in which the wave travels. An example of this type of wave is a mechanical wave projected along a tight string. The string moves at right angles to the wave motion. Electromagnetic waves are another example of transverse waves. The directions of the electric and magnetic fields are perpendicular to the wave motion. In longitudinal waves the disturbance takes place parallel to the wave motion. A longitudinal wave consists of a series of compressions and rarefactions (states of maximum and minimum density and pressure, respectively). Such waves are always mechanical in nature and thus require a medium through which to travel. Sound waves are an example of longitudinal waves. Waves that result from a stone being dropped into water appear as a series of circles. These are called circular waves and can be generated in a ripple tank for study. Waves on water that appear as a series of parallel lines are called plane waves.

characteristics of waves
All waves have a wavelength. This is measured as the distance between successive crests (or successive troughs) of the wave. It is given the Greek symbol λ. The frequency of a wave is the number of vibrations per second. The reciprocal of this is the wave period. This is the time taken for one complete cycle of the wave oscillation. The speed of the wave is measured by multiplying wave frequency by the wavelength.

properties of waves
When a wave moves from one medium to another (for example a light wave moving from air to glass) it moves with a different speed in the second medium. This change in speed causes it to change direction. This property is called refraction. The angle of refraction depends on whether the wave is speeding up or slowing down as it changes medium. Reflection occurs whenever a wave hits a barrier. The wave is sent back, or reflected, into the medium at a different angle. The angle of incidence (the angle between the ray and a perpendicular line drawn to the surface) is equal to the angle of reflection (the angle between the reflected ray and a perpendicular to the surface). See also total internal reflection. An echo is the repetition of a sound wave by reflection from a surface. All waves spread slightly as they travel. This is called diffraction and it occurs chiefly when a wave interacts with a solid object. The degree of diffraction depends on the relationship between the wavelength and the size of the object (or gap through which the wave travels). If the two are similar in size, diffraction occurs

and the wave can be seen to spread out. Large objects cast shadows because the difference between their size and the wavelength is so large that light waves are not diffracted round the object. A dark shadow results. When two or more waves meet at a point, they interact and combine to produce a resultant wave of larger or smaller amplitude (depending on whether the combining waves are in or out of phase with each other). This is called interference. Transverse waves can exhibit polarization. If the oscillations of the wave take place in many different directions (all at right angles to the directions of the wave) the wave is unpolarized. If the oscillations occur in one plane only, the wave is polarized. Light, which consists of transverse waves, can be polarized.

the amplitude of oscillations
The amplitude is the maximum displacement of an oscillation from the equilibrium position (the height of a crest or the depth of a trough). With a sound wave, for example, amplitude corresponds to the intensity (loudness) of the sound. If a mechanical system is made to vibrate by applying oscillations to it, the system vibrates. As the frequency of the oscillations is varied, the amplitude of the vibrations reaches a maximum at the natural frequency of the system. If a force with a frequency equal to the natural frequency is applied, the vibrations can become violent, a phenomenon known as resonance.

electromagnetic waves
All electromagnetic waves travel at the same speed, the speed of light. They vary in wavelength and frequency to give the broad range of waves found in the electromagnetic spectrum. They include radio waves, infrared radiation, visible light, ultraviolet radiation, X-rays, and gamma rays. Many of the colours observed in everyday objects are seen because of the phenomenon of the absorption of light. Rainbows display the colours of the spectrum and are formed by the refraction and reflection of the Sun's rays through rain or mist.

wavelength
distance between successive crests of a wave. The wavelength of a light wave determines its colour; red light has a wavelength of about 700 nanometres, for example. The complete range of wavelengths of electromagnetic waves is called the electromagnetic spectrum.

weak nuclear force or weak interaction
one of the four fundamental forces of nature, the other three being the gravitational force or gravity, the electromagnetic force, and the strong nuclear force. It causes radioactive beta decay and other subatomic reactions. The particles that carry the weak force are called weakons (or intermediate vector bosons) and comprise the positively and negatively charged W particles and the neutral Z particle.

weakon
see intermediate vector boson.

weber

SI unit (symbol Wb) of magnetic flux (the magnetic field strength multiplied by the area through which the field passes). It is named after German chemist Wilhelm Weber. One weber equals 10^8 maxwells.

A change of flux at a uniform rate of one weber per second in an electrical coil with one turn produces an electromotive force of one volt in the coil.

Wien's displacement law

in physics, a law of radiation stating that the wavelength carrying the maximum energy is inversely proportional to the absolute temperature of a black body: the hotter a body is, the shorter the wavelength. It has the form $\lambda_{max}T$ = constant, where λ_{max} is the wavelength of maximum intensity and T is the temperature. The law is named after German physicist Wilhelm Wien.

work

measure of the result of transferring energy from one system to another to cause an object to move. Work should not be confused with energy (the capacity to do work, which is also measured in joules) or with power (the rate of doing work, measured in joules per second).

Work is equal to the product of the force used and the distance moved by the object in the direction of that force. If the force is F newtons and the distance moved is d metres, then the work W is given by: $W=Fd$. For example, the work done when a force of 10 newtons moves an object 5 metres against resistance is 50 joules (50 newton-metres).

W particle

elementary particle, one of the intermediate vector bosons responsible for transmitting the weak nuclear force. The W particle exists as both W^+ and W^-.

X-ray

band of electromagnetic radiation in the wavelength range 10^{-11} to 10^{-9} m (between gamma rays and ultraviolet radiation; see electromagnetic waves). Applications of X-rays make use of their short wavelength (as in X-ray diffraction) or their penetrating power (as in medical X-rays of internal body tissues). X-rays are dangerous and can cause cancer.

X-rays with short wavelengths pass through most body tissues, although dense areas such as bone prevent their passage, showing up as white areas on X-ray photographs. The X-rays used in radiotherapy have very short wavelengths that penetrate tissues deeply and destroy them. X-rays were discovered by German experimental physicist Wilhelm Röntgen in 1895 and formerly called roentgen rays. They are produced when high-energy electrons from a heated filament cathode strike the surface of a target (usually made of tungsten) on the face of a massive heat-conducting anode, between which a high alternating voltage (about 100 kV) is applied.

Z particle

elementary particle, one of the intermediate vector bosons responsible for carrying the weak nuclear force. The Z particle is neutral.

Appendix

Nobel Prize for Physics

Year	Winner(s)[1]	Awarded for
1901	Wilhelm Röntgen (Germany)	discovery of X-rays
1902	Hendrik Lorentz (Netherlands) and Pieter Zeeman (Netherlands)	influence of magnetism on radiation phenomena
1903	Henri Becquerel (France)	discovery of spontaneous radioactivity
	Pierre Curie (France) and Marie Curie (France)	research on radiation phenomena
1904	John Strutt (Lord Rayleigh, UK)	densities of gases and discovery of argon
1905	Philipp von Lenard (Germany)	work on cathode rays
1906	Joseph J Thomson (UK)	theoretical and experimental work on the conduction of electricity by gases
1907	Albert Michelson (USA)	measurement of the speed of light through the design and application of precise optical instruments such as the interferometer
1908	Gabriel Lippmann (France)	photographic reproduction of colours by interference
1909	Guglielmo Marconi (Italy) and Karl Ferdinand Braun (Germany)	development of wireless telegraphy
1910	Johannes van der Waals (Netherlands)	equation describing the physical behaviour of gases and liquids
1911	Wilhelm Wien (Germany)	laws governing radiation of heat

Year	Winner(s)[1]	Awarded for
1912	Nils Dalén (Sweden)	invention of light-controlled valves, which allow lighthouses and buoys to operate automatically
1913	Heike Kamerlingh Onnes (Netherlands)	studies of properties of matter at low temperatures
1914	Max von Laue (Germany)	discovery of diffraction of X-rays by crystals
1915	William Bragg (UK) and Lawrence Bragg (UK)	X-ray analysis of crystal structures
1916		no award
1917	Charles Barkla (UK)	discovery of characteristic X-ray emission of the elements
1918	Max Planck (Germany)	formulation of quantum theory
1919	Johannes Stark (Germany)	discovery of Doppler effect in rays of positive ions, and splitting of spectral lines in electric fields
1920	Charles Guillaume (Switzerland)	discovery of anomalies in nickel-steel alloys
1921	Albert Einstein (Switzerland)	theoretical physics, especially law of photoelectric effect
1922	Niels Bohr (Denmark)	discovery of the structure of atoms and radiation emanating from them
1923	Robert Millikan (USA)	discovery of the electric charge of an electron, and study of the photoelectric effect
1924	Karl Siegbahn (Sweden)	X-ray spectroscopy
1925	James Franck (Germany) and Gustav Hertz (Germany)	discovery of laws governing the impact of an electron upon an atom
1926	Jean Perrin (France)	confirmation of the discontinuous structure of matter
1927	Arthur Compton (USA)	transfer of energy from electromagnetic radiation to a particle

Year	Winner(s)[1]	Awarded for
	Charles Wilson (UK)	invention of the Wilson cloud chamber, by which the movement of electrically charged particles may be tracked
1928	Owen Richardson (UK)	work on thermionic phenomena and associated law
1929	Louis Victor de Broglie (France)	discovery of the wavelike nature of electrons
1930	Chandrasekhara Raman (India)	discovery of the scattering of single-wavelength light when it is passed through a transparent substance
1931	no award	
1932	Werner Heisenberg (Germany)	creation of quantum mechanics
1933	Erwin Schrödinger (Austria) and Paul Dirac (UK)	development of quantum mechanics
1934	no award	
1935	James Chadwick (UK)	discovery of the neutron
1936	Victor Hess (Austria)	discovery of cosmic radiation
	Carl Anderson (USA)	discovery of the positron
1937	Clinton Davisson (USA) and George Thomson (UK)	diffraction of electrons by crystals
1938	Enrico Fermi (Italy)	use of neutron irradiation to produce new elements, and discovery of nuclear reactions induced by slow neutrons
1939	Ernest Lawrence (USA)	invention and development of the cyclotron, and production of artificial radioactive elements
1940	no award	
1941	no award	
1942	no award	
1943	Otto Stern (USA)	molecular-ray method of investigating elementary particles, and discovery of magnetic moment of proton

Year	Winner(s)[1]	Awarded for
1944	Isidor Isaac Rabi (USA)	resonance method of recording the magnetic properties of atomic nuclei
1945	Wolfgang Pauli (Austria)	discovery of the exclusion principle
1946	Percy Bridgman (USA)	development of high-pressure physics
1947	Edward Appleton (UK)	physics of the upper atmosphere
1948	Patrick Blackett (UK)	application of the Wilson cloud chamber to nuclear physics and cosmic radiation
1949	Hideki Yukawa (Japan)	theoretical work predicting existence of mesons
1950	Cecil Powell (UK)	use of photographic emulsion to study nuclear processes, and discovery of pions (pi mesons)
1951	John Cockcroft (UK) and Ernest Walton (Ireland)	transmutation of atomic nuclei by means of accelerated subatomic particles
1952	Felix Bloch (USA) and Edward Purcell (USA)	precise nuclear magnetic measurements
1953	Frits Zernike (Netherlands)	invention of phase-contrast microscope
1954	Max Born (UK)	statistical interpretation of wave function in quantum mechanics
	Walther Bothe (West Germany)	coincidence method of detecting the emission of electrons
1955	Willis Lamb (USA)	structure of hydrogen spectrum
	Polykarp Kusch (USA)	determination of magnetic moment of the electron
1956	William Shockley (USA), John Bardeen (USA), and Walter Houser Brattain (USA)	study of semiconductors, and discovery of the transistor effect
1957	Tsung-Dao Lee (China) and Chen Ning Yang (China)	investigations of weak interactions between elementary particles
1958	Pavel Cherenkov (USSR), Ilya Frank (USSR), and Igor Tamm (USSR)	discovery and interpretation of Cherenkov radiation

Year	Winner(s)[1]	Awarded for
1959	Emilio Segrè (USA) and Owen Chamberlain (USA)	discovery of the antiproton
1960	Donald Glaser (USA)	invention of the bubble chamber
1961	Robert Hofstadter (USA)	scattering of electrons in atomic nuclei, and structure of protons and neutrons
	Rudolf Mössbauer (West Germany)	resonance absorption of gamma radiation
1962	Lev Landau (USSR)	theories of condensed matter, especially liquid helium
1963	Eugene Wigner (USA)	discovery and application of symmetry principles in atomic physics
	Maria Goeppert-Mayer (USA) and Hans Jensen (Germany)	discovery of the shell-like structure of atomic nuclei
1964	Charles Townes (USA), Nikolai Basov (USSR), and Aleksandr Prokhorov (USSR)	work on quantum electronics leading to construction of oscillators and amplifiers based on maser–laser principle
1965	Sin-Itiro Tomonaga (Japan), Julian Schwinger (USA), and Richard Feynman (USA)	basic principles of quantum electrodynamics
1966	Alfred Kastler (France)	development of optical pumping, whereby atoms are raised to higher energy levels by illumination
1967	Hans Bethe (USA)	theory of nuclear reactions, and discoveries concerning production of energy in stars
1968	Luis Alvarez (USA)	elementary-particle physics, and discovery of resonance states, using hydrogen bubble chamber and data analysis
1969	Murray Gell-Mann (USA)	classification of elementary particles, and study of their interactions
1970	Hannes Alfvén (Sweden)	work in magnetohydrodynamics and its applications in plasma physics

Year	Winner(s)[1]	Awarded for
	Louis Néel (France)	work in antiferromagnetism and ferromagnetism in solid-state physics
1971	Dennis Gabor (UK)	invention and development of holography
1972	John Bardeen (USA), Leon Cooper (USA), and John Robert Schrieffer (USA)	theory of superconductivity
1973	Leo Esaki (Japan) and Ivar Giaever (USA)	tunnelling phenomena in semiconductors and superconductors
	Brian Josephson (UK)	theoretical predictions of the properties of a supercurrent through a tunnel barrier
1974	Martin Ryle (UK) and Antony Hewish (UK)	development of radioastronomy, particularly the aperture-synthesis technique, and the discovery of pulsars
1975	Aage Bohr (Denmark), Ben Mottelson (Denmark), and James Rainwater (USA)	discovery of connection between collective motion and particle motion in atomic nuclei, and development of theory of nuclear structure
1976	Burton Richter (USA) and Samuel Ting (USA)	discovery of the psi meson
1977	Philip Anderson (USA), Nevill Mott (UK), and John Van Vleck (USA)	contributions to understanding electronic structure of magnetic and disordered systems
1978	Peter Kapitza (USSR)	invention and application of low-temperature physics
	Arno Penzias (USA) and Robert Wilson (USA)	discovery of cosmic background radiation
1979	Sheldon Glashow (USA), Abdus Salam (Pakistan), and Steven Weinberg (USA)	unified theory of weak and electromagnetic fundamental forces, and prediction of the existence of the weak neutral current
1980	James W Cronin (USA) and Val Fitch (USA)	violations of fundamental symmetry principles in the decay of neutral kaon mesons

Year	Winner(s)[1]	Awarded for
1981	Nicolaas Bloembergen (USA) and Arthur Schawlow (USA)	development of laser spectroscopy
	Kai Siegbahn (Sweden)	high-resolution electron spectroscopy
1982	Kenneth Wilson (USA)	theory for critical phenomena in connection with phase transitions
1983	Subrahmanyan Chandrasekhar (USA)	theoretical studies of physical processes in connection with structure and evolution of stars
	William Fowler (USA)	nuclear reactions involved in the formation of chemical elements in the universe
1984	Carlo Rubbia (Italy) and Simon van der Meer (Netherlands)	contributions to the discovery of the W and Z particles (weakons)
1985	Klaus von Klitzing (West Germany)	discovery of the quantized Hall effect
1986	Ernst Ruska (West Germany)	electron optics, and design of the first electron microscope
	Gerd Binnig (West Germany) and Heinrich Rohrer (Switzerland)	design of scanning tunnelling microscope
1987	Georg Bednorz (West Germany) and Alex Müller (Switzerland)	superconductivity in ceramic materials
1988	Leon M Lederman (USA), Melvin Schwartz (USA), and Jack Steinberger (USA)	neutrino-beam method, and demonstration of the doublet structure of leptons through discovery of muon neutrino
1989	Norman Ramsey (USA)	measurement techniques leading to discovery of caesium atomic clock
	Hans Dehmelt (USA) and Wolfgang Paul (Germany)	ion-trap method for isolating single atoms
1990	Jerome Friedman (USA), Henry Kendall (USA), and Richard Taylor (Canada)	experiments demonstrating that protons and neutrons are made up of quarks

Year	Winner(s)[1]	Awarded for
1991	Pierre-Gilles de Gennes (France)	work on disordered systems including polymers and liquid crystals; development of mathematical methods for studying the behaviour of molecules in a liquid on the verge of solidifying
1992	Georges Charpak (France)	invention and development of detectors used in high-energy physics
1993	Joseph Taylor (USA) and Russell Hulse (USA)	discovery of first binary pulsar (confirming the existence of gravitational waves)
1994	Clifford Shull (USA) and Bertram Brockhouse (Canada)	development of technique known as 'neutron scattering' which led to advances in semiconductor technology
1995	Frederick Reines (USA)	discovery of the neutrino
	Martin Perl (USA)	discovery of the tau lepton
1996	David Lee (USA), Douglas Osheroff (USA), and Robert Richardson (USA)	discovery of superfluidity in helium-3
1997	Claude Cohen-Tannoudji (France), William Phillips (USA), and Steven Chu (USA)	discovery of a way to slow down individual atoms using lasers for study in a near-vacuum
1998	Robert B Laughlin (USA), Horst L Störmer (USA), and Daniel C Tsui (USA)	discovery of a new form of quantum fluid with fractionally charged excitations
1999	Gerardus 'T Hooft (the Netherlands) and Martinus Veltman (the Netherlands)	elucidating the quantum structure of electroweak interactions in physics
2000	Zhores I Alferov (Russia) and Herbert Kroemer (Germany)	development of semiconductor heterostructures, which lead to faster transistors and more efficient laser diodes
	Jack St Clair Kilby (USA)	co-invention of the integrated circuit

[1] Nationality given is the citizenship of recipient at the time award was made.

Index

Note: Page numbers in *Italics* refer to illustrations

Abdus Salam Centre for Theoretical
    Physics (ICTP), 105
Accelerator Physics Group Web site, 129
African Academy of Sciences, 105
ALEPH Experiment Web site, 129
Alfvén, Hannes Olof Gösta, 13
    biographical sketch, 49
Alpha radiation, 7
Alpher-Bethe-Gamow hypothesis, 73
Alvarez, Luis Walter, biographical
    sketch, 49–50
American Association for the Advance-
    ment of Science (AAAS), 105–6
American Physical Society, 106
Ames Research Center, 106
Anderson, Carl David, 11, 36, 37
    biographical sketch, 50–1, *51*
Andromeda, 20
Animated Holographer Web site, 129
Antarctic Muon and Neutrino Detector
    Array (AMANDA), 46
Anti-electrons, 11
Antimatter, 11–12, 23, 46
Antineutrons, 11, 40
Antiparticles, 65
Antiprotons, 11, 94
Appleton, Edward Victor, biographical
    sketch, 51–2
Appleton layer (F layer), 38, 51
Astatine, 37, 94
Aston, Francis William, biographical
    sketch, 52
Astrophysics, work of George Gamow,
    72–3
Atmosphere
    Appleton layer (F layer), 51
    conducting layer, 25
    ionized layer, 34
    Kennelly-Heaviside layer (E layer), 51
Atmospheric pressure, 29
Atomic bomb, 19, *20*, 49, 50, 56, 61, 71,
    101
    work of Enrico Fermi, 69
    work of Ernest Lawrence, 82
    work of J Robert Oppenheimer, 87
    work of Rudolf Peierls, 88
Atomic chain reaction, 38, 88
Atomic energy, 17–19
Atomic fission, 36
Atomic numbers, work of Henry
    Moseley, 85–6

Atomic Physics and Human Knowledge
    (Bohr), 57
Atomic pile, 69
Atomic structure, 8–10
*Atomic Theory and the Description of
    Nature* (Bohr), 57
Atomic transmutation, 55
Atomic weights, Web site, 137
Atoms, 2, 7–8, 26, 28
    splitting, 32, 34
    work of John Cockcroft, 60
    work of Ernest Walton, 98
    Web site, 139
    work of Niels Bohr, 55
'Audion tube', 28
Automatic sun valve, 29

Balmer, Johann, 11
Balmer series, 11
Bardeen, John, 13, 38, *39*
    biographical sketch, 52–3
Barium, 43
Barkla, Charles Glover, biographical
    sketch, 53
BBC Science in Action Web site, 129
BCS theory, 13–14
Beam Me Up: Photons and
    Teleportation Web Site, 129
Becquerel, Henri, 7, 25
Bednorz, Johannes Georg, 14
    biographical sketch, 53
Bell Burnell, (Susan) Jocelyn,
    biographical sketch, 54
Bell, John, 12
    biographical sketch, 54
Bell Research Labs, AT&T, 107
Berkeley Lab, 107
Beta decay, 14
Beta particles, 25
Beta radiation, 7
Beta rays, 63
'Big Bang' theory, 21–2, 38, 42, 72
    hot Big Bang, 73
Biographies of Physicists Web site, 130
Black holes, 21–3, *22*, 42
    work of Stephen Hawking, 76, 78
Blackett, Patrick Maynard Stuart
    biographical sketch, 54–5
    Web site, 130
Body scanner, 14
Bohr, Aage, *56*

Bohr, Niels Henrik David, 8, 10–11, 12, 17, 30, 31, 32
  biographical sketch, 55–7, *56*
Born, Max
  biographical sketch, 57
  Web site, 130
Bose-Einstein condensate, 46
Bosons, 16, 24, 100
Bragg, William Henry, 6
  biographical sketch, 58
Bragg, (William) Lawrence, 6
  biographical sketch, 57–8
Bragg's law, 58
Brattain, Walter, 38, *39*
Bridgman, Percy William, biographical sketch, 58–9
A Brief History of Time (Hawking), 76, 125
Broglie, Louis Victor Pierre Raymond de, 11
  biographical sketch, 59
  Web site, 131
Brookhaven National Laboratory, 107
Brout, Robert, 24
Brownian motion, 3, 28
  work of Albert Einstein, 66
Bubble chamber, 39, *40*, 50

Caesium, 43
California Institute of Technology, 107–8
The Cambridge Guide to the Material World (Cotterill), 124
Carbon-14 dating, 38
Cathode rays, 2
The Cause and Nature of Radioactivity (Rutherford and Soddy), 25
Cavendish Laboratory, 108
CERN (laboratory for nuclear research), 39, 42, 44, 57, 108
  Web site, 133
Chadwick, James, 10, 51, 123
  biographical sketch, 59–60
Challenger space shuttle disaster, 71
The Character of Physical Law (Feynman), 124
Charmed quark, 42, 43
  work of Sheldon Glashow, 74, 75
Chemical bonding, 34
Chemical properties of materials, 13
'Cherenkov radiation', 36
Chernobyl, 19
Clerk Maxwell, James *see* Maxwell
Cloud chamber, 30, 32, 50, 55, 62
Cockcroft, John Douglas, 15, *61*
  biographical sketch, 60
Colour confinement, 16
Colour Matters Web site, 130

Complementarity, work of Niels Bohr, 55
Compton, Arthur Holly, 6
  biographical sketch, 60–62
Compton effect, 6, 33, 60
Cooper, Leon Niels, 13
  biographical sketch, 62–3
Cooper pairs, 62
Coppler, Christian Johann, 21
Cornell University, 109
Correspondence principle, 32
Cosmic radiation, work of Patrick Blackett, 54
Cosmic rays, 85
Cosmology, 20
CP (charge parity) nonconservation, 41
Crommelin, Andrew, 5
Cryogenics Web site, 131
Crystals
  work of Lawrence Bragg, 57–8
  work of Max Born, 57
  work of William Bragg, 58
  and X-rays, 82
Curie, Marie, 8, 26, *27*, 29
  biographical sketch, 63–4
Curie, Marie and Pierre, Web site, 133
Curie, Pierre, 8, *27*
Cyclotrons, 15, 35, *35*, 37
  'Nimrod', 41
  work of Ernest Lawrence, 82

Daresbury Laboratory, 109
Davisson, Clinton Joseph, 11
  biographical sketch, 64
de Broglie *see* Broglie
De Forest, Lee, biographical sketch, 64–5
Debierne, André-Louis, 26
Debye and Strong Electrolytes Web site, 131
Deuterium (heavy hydrogen), 19
  discovery, 35
Deutsches Elektronen-Synchrontron (DESY), 109
Dinosaurs, extinction, work of Luis Alvarez, 49–50
Diode valve, 26
Dirac, Paul Adrien Maurice, 11
  biographical sketch, 65
DNA, work of George Gamow, 73
DNA structure, 6
Does God Play Dice? (Stewart), 126
Domain theory of ferro-magnetism, 28
Doppler effect, 21
  Web site, 131
Dreams of a Final Theory (Lederman and Aschramm), 125
Dreams of a Final Theory (Weinberg), 100

Eddington, Arthur, 5, 32
Ehrenfest, Paul, *3*
Eightfold Way, 41
Einstein, Albert, 3, *3*, 6, 18, 28, 29, 30,
    31, 35
  biographical sketch, 65–8
  Brownian motion, 66
  general theory of relativity, 4–5, 31,
    67–8
  law of photochemical equivalence, 31
  photoelectricity, 66
  special theory of relativity, 4, 28, 66–7
  Web sites, 132, 138
Einstein for Beginners (Schwartz and
    McGuiness), 126
Einstein (Bernstein), 123
Einstein's Mirror (Hey and Walters), 125
Einstein's Universe (Calder), 123
Electricity, 1
Electromagnetic radiation, 3–4, 5–6, 29
Electromagnetism, 1, 2
  work of Abdus Salam, 91
  work of Hendrik Lorentz, 82–3
Electronic charge, work of Robert
    Millikan, 85
Electrons, 6, 7, 11
  diffraction, 34
  oil-drop experiment, 30
  tunnelling, 40
  wave nature, 64
  work of J J Thomson, 97–8
Electroscope, 26
Electroweak theory, 17, 42, 99–100
Emilio Segrè Visual Archives Web site,
    132
Energy
  concept, 1–2
  from atoms, 17–19
EnergyEd Web site, 132
Englert, François, 24
Eric's Treasure Trove of Physics Web
    site, 132
Esaki diodes, 68
Esaki, Leo, biographical sketch, 68–9
*Essays 1968–1962 on Atomic Physics
    and Human Knowledge* (Bohr),
    57
European Physical Society, 110
European Synchrotron Radiation
    Facility (EFRF), 45
The Evolution of Physics (Einstein and
    Infeld), 1124
Exclusion principle, work of Wolfgang
    Pauli, 87–8
ExploreScience.com Web site, 133

Faraday, Michael, 1
Fermi, Enrico, 10, 19, 36, 69–70

Fermi National Accelerator Laboratory
    (Fermilab), 42, 110
  Web site, 135
Fermi–Dirac statistics, 65
Feynman diagrams, 70
The Feynman Lectures on Physics
    (Feynman), 71
Feynman, Richard, 15
  biographical sketch, 70
  Web site, 134
Fibre Optic Chronology Web site, 134
The First Three Minutes (Weinberg), 100
Fission
  of uranium, 18, 71–2
  work of Leo Szilard, 97
  work of Lise Meitner, 84–5
Fleming, John, 26
The Forces of Nature (Close, Marten and
    Sutton), 123
*The Foundation of the General Theory
    of Relativity* (Einstein), 31
Franck, James, biographical sketch, 71
Franck–Condon principle, 71
Franklin, Rosalind, 6
Free Fall and Air Resistance Web site,
    134
Frisch, Otto Robert, 17, 19
  biographical sketch, 71
From Falling Bodies to Radio Waves
    (Segrè), 126
From Nuclei to Stars (Molinard and
    Ricci), 126
From Quarks to the Cosmos (Carrigan),
    123
Frozen-in-flux theorem, 49
Fusion energy, 19
Fusion Web site, 134

Gabor, Dennis, 7
  biographical sketch, 72
Gamma photons, 22
Gamma radiation, 8
Gamma rays, 25, 26
Gamow, George, 17, 21
  biographical sketch, 72–3
Gaseous diffusion, 37
Gases, 13
Gauge bosons, 16
GCSE Bitesize Physics Web site, 134–5
Geiger counter, 29, 73
Geiger, Hans, 8, 29
  biographical sketch, 73–4
Gell-Mann, Murray, 16, 41, 42, 125
  biographical sketch, 73–4
General theory of relativity, 4–5, 31–2,
    67–8
Genius (Gleick), 124
Germer, Lester, 11

Glashow, Sheldon Lee, 17, 43
biographical sketch, 74–5
Gluons, 16, 17, 42, 43, 45
Grand unified theory (GUT),
24, 43
Gravitation, 1, 5
Gravitation and Cosmology (Weinberg),
100
Gravitational field theory, 76
Gravitational lensing, 5
Gravitons, 17
Gravity, 22, 32
Web sites, 133, 134
Gravity waves, 23
Great Nebula, 20

Hadrons, 16
Hahn, Otto, 17, *18*
biographical sketch, 74–5
Harvard-Smithsonian Center for
Astrophysics, 110–11
Hassium, 44
Hawking radiation, 76
Hawking, Stephen, 42, 77
biographical sketch, 76
Web site, 135
Heaviside, Oliver, Web site, 135
Heisenberg, Werner, 12, 34
biographical sketch, 78
Web site, 135
Helium, 13, 26
cooling, 36
liquefaction, 29
solidification, 34
superfluidity, 37, 70, 80–81
Hertz, Gustav, biographical sketch,
79
Hertz, Heinrich, 1
Heterodyne principle, 25
Higgs boson, 24, 78
Higgs, Peter Ware, 24
biographical sketch, 79
High pressure physics, work of Percy
Bridgman, 58–9
Hiroshima, 19
Holography, 7, 38
Web site, 129
work of Dennis Gabor, 72
How Things Work: the Physics of
Everyday Life Web site, 136
How Things Work Web site, 136
How to Make a Pinhole Camera Web
site, 136
Hubble, Edwin, 20–21
Hydrogen, 31
Hydrogen bomb, 19
Hydrogen spectrum, 11
Hyperspace (Kaku), 125

Infrared (IV) radiation, 7
Institut de Génie Atomique (IGA), 111
Institute for Advanced Study, 111–12
Institute of Fluid Mechanics, Göttingen,
112
Institute of Physics, 113
Institute for Plasma Research, 112
Interactive Physics Problem Set Web
site, 136
International Association of
Mathematical Physics, 113
International Bureau of Weights and
Measures, 113
Isotopes, 30, 31
and geological age, 32
Web site, 137
Isotopes (Aston), 52

J particle, 98
Jeans, Sir James Hopwood, Web site, 137
Jet Propulsion Laboratory, 114
Joint European Torus (JET), 44
Joint Institute for Nuclear Research
(JINR), 114
Josephson, Brian David, biographical
sketch, 79–80
Josephson effect, 41, 80

Kaleidoscope Heaven Web site, 137
Kapitza, Peter Leonidovich (Pyotr
Kapitsa), 13, 37, *81*
biographical sketch, 80
Kennelly-Heaviside layer (E layer), 51
Kharlov Institute of Physics and
Technology (KIPT), 114–15

Laboratory of Molecular Biology, 115
Laboratory of Molecular Biophysics,
115
Landau, Lev Davidovich, Web site, 138
The Large Scale Structure of Space-Time
(Hawking and Ellis), 76
Lasers, 7
work of Theodore Maiman, 83–4
Lasers (Harbison and Nahory), 125
Laue, Max Theodor Felix von,
biographical sketch, 80–1
Lawrence, Ernest O, 15, *35*
biographical sketch, 82
Lawrence Livermore National
Laboratory, 116
Learn Physics Today Web site, 138
Lemaître, Georges, 21
LEP (Large Electron-Positron Collider),
15–16, 23
Leptons, 16, 46
Leptoquarks Web site, 138
LHC (Large Hadron Collider), 23

Light
  measurement of speed, 33, 34
  slowing down, 47
  velocity, 38
Light! Web site, 139
Line radiation, 39
Linear accelerator, 50
Linear Accelerator Center (SLAC), 116
Little Shop of Physics: Online
  Experiments Web site, 139
Lorentz, Hendrik Antoon,
  biographical sketch, 82–3
Los Alamos National Laboratory,
  116–17
Los Alamos Web site, 139
Low temperature physics, 13
  and magnetism, 80

McMillan, Edwin Mattison,
  biographical sketch, 83
Magnetic resonance accelerators, 37
Magnetism, 1, 28, 38, 80
  Web sites, 131, 132
Magnetohydrodynamics, 49
Maiman, Theodore Harold, 7
  biographical sketch, 83
Manhattan Project, 19, 60, 69, 87, 96
Marconi, Guglielmo, 7, 25
Marsden, Ernest, 8
Mass spectrometer, 31, 52
Mass Spectrometry Web site, 137
Massachusetts Institute of Technology,
  117
Maths and Physics Help Page Web site,
  139
Matrix mechanics, 57, 78
Matter
  framework of, 12–14
  structure of, 2
Max Planck Society for the
  Advancement of Science, 117
Maxwell, James Clerk, 1, 3
Maxwell's equations, 3–4
The Meaning and Structure of Physics
  (Cooper), 63
Mechanical Properties of Plastics Web
  site, 140
Medical Research Council Laboratory,
  Cambridge, 118
Meissner effect, 36
Meitner, Lise, 17, *18*
  biographical sketch, 84
Mesons, 22, 36
  work of Hideki Yukawa, 101
Mesotron (Meson muon), 50
Metals, properties, 13
Michelson, Albert A, 2
Michelson–Morley experiment, 67

Mictrotron, 50
Millikan, Robert Andrews, biographical
  sketch, 84–5
Molecular vacuum pump, 29
Morley, Edward W, 2
Moseley, Henry Gwyn Jeffreys, 8
  biographical sketch, *85*, 85–6
Moseley's law, 85
Müller, Karl Alexander, 14
  biographical sketch, 86
Munsell colour scheme, 31
Muons, 14, 37, 50, 101
The Mystery of the Quantum World
  (Squires), 126

Nagasaki, 19, *20*
National Academy of Sciences, 118
National Institution for Standards and
  Technology (NIST), 118
National Physical Laboratory (NPL),
  119
National Science Foundation (NSF),
  119
National Society of Black Physicists
  (NSBP), 119–20
The Nature of Space and Time
  (Hawking and Penrose), 76
Neon lighting, 29
Neptunium, 37, 83
NetScience: Physics Web site, 140
Neutral X-particle, 41
Neutrinos, 14, 40, 42, 75, 87, 88
  work of Enrico Fermi, 70
The Neutron and the Bomb: A
  Biography of Sir James Chadwick
  (Brown), 123
Neutron-proton system, 88
Neutrons, 10, 22, 36, 40
  work of James Chadwick, 59
Newton, Isaac, 1
Next Generation Space Telescope
  (NGST), 23
Niels Bohr Institute, 120
  Web site, 140
No Ordinary Genius (Sykes), 127
Nobel Prize for Physics, winners
  (1901–2000), 263–70
Novosibirsk State University Physics
  Department, 120
Nuclear atom, concept, 30, 31
Nuclear energy, 17, 19, 76
  Web site, 140
Nuclear fission, 76
Nuclear fusion, 44
  cold fusion, 45
Nuclear physics, parity or asymmetry,
  100
Nuclear Physics Web site, 140

Nuclear reactions, work of Enrico
    Fermi, 69–70
Nuclear reactor, 19, 36
Nucleus, 8–10
    liquid-droplet model, 55–6

Omega-minus particle, 41
Onnes, Heike Kamerlingh, 13, 30
OPAL detector, *15*
Oppenheimer, J Robert, biographical
    sketch, 86–7
Optical parametric oscillator, 42
Orbitals, 12–13
Order, Chaos, Order (Stehle), 126

Parabola spectrograph, 31
Parity in nuclear physics, 100
Particle accelerator, 15, 35, 36, 60, *99*
Particle Adventure Web site, 140
The Particle Explosion (Close), 123
Particle Physics Web site, 141
Particles and Forces (Davies), 124
Pauli, Wolfgang, 14
    biographical sketch, 87
    Web site, 141
Pearl Harbor, 19
Peierls, Rudolf Ernst, 19
    biographical sketch, 87–8
Phase stability principle, 37
Phase transitions, 42
Photochemical equivalence law, 31
Photoelectric cells, 25
Photoelectric law, 65
Photoelectricity, work of Albert
    Einstein, 66
Photons, 6, 17, 62, 66, 79
    Web site, 130
The Physicists (Snow), 126
Physics 2000 Web site, 141
Physics of High Pressure (Bridgman), 59
Physics for the Inquiring Mind (Rogers),
    126
Physics Zone Web site, 141
Physics-related information Web site,
    141
PhysicsTutor.com Web site, 141
Pi-meson (Pion), 14, 16, 38, 101
Pilot waves, 11
Pions, 14, 16, 38, 101
Pitchblende, 8, 63
Planck, Max, 5, 25
    Web site, 142
Planck, Max Karl Ernst Ludwig, 5, 25
biographical sketch, 88–9
    Web site, 142
Planck's constant, 89
    work of Robert Millikan, 85
Planck's theory, 6

Plasma, 13, 19
    Web site, 136
    work of Hannes Alfvén, 49
Plastics, Web site, 140
Plutonium, 19, 37, 38
Polonium, 63
Positron, 11, 36, 50
Protactinium, 76, 84
Protons, 10, 22, 29, 32, 40
Pugwash movement, 49
Pulsars, 54

Q is for Quantum (Kane), 125
*QED: The strange theory of light and
    matter* (Feynman), 124
Quanta, 55
Quantum chromodynamics (QCM), 42
Quantum electrodynamics, 15, 93
    work of Richard Feynman, 70
Quantum mechanics, 57, 78, 93
    work of Paul Dirac, 65
Quantum physics, 10–11, 25
The Quantum Self (Zohar), 127
The Quantum Society (Zohar and
    Marshall), 127
Quantum theory, 12
    Web site, 142
    work of John Bell, 54
    work of Niels Bohr, 55
    work of Max Born, 57
    work of James Franck, 71
    work of Gustav Hertz, 79
    work of Werner Heisenberg, 78
    work of Robert Millikan, 85
    work of Max Planck, 88–9
    work of Erwin Schrödinger, 92
Quark theory, work of George Zweig,
    102
Quarks, 16, 22, 41, 46
    charm (charmed), 42, 43
    work of Sheldon Glashow, 74, 75
    sixth discovered, 44
    Web site, 136
    work of Murray Gell-Mann, 74
Quarks (Fritsch), 124
Quarks (Weinberg), 127
Quasars, 5, 21
'Quasiparticles', 46
The Quest for Absolute Zero
    (Mendelssohn), 125
Quintessence, 23, 47

Radar, 7, 26
    Web site, 135
Radiation, 5–6, 7–8, 31
    alpha, 7
    applications, 6–7
    beta, 7

cosmic, 54, 85
electromagnetic, 3–4, 5–6, 29
gamma, 8
infrared (IV), 7
line, 39
ultraviolet (UV), 6–7
Web sites, 142–3
Radio waves, 7, 25, 51–2
Radio-activity (Rutherford), 26
Radioactive decay, work of George
Gamow, 73
Radioactivity, 14, 25–6, 29
Web sites, 142–3
work of Marie Curie, 63
work of Lise Meitner, 84
work of Ernest Rutherford, 90–91
Radium, 8, 26, 31
medical use, 25
work of Marie Curie, 63
Ramsay, William, 26
Red shift, 20–21, *22*
The Refrigerator and the Universe
(Goldstein and Goldstein), 124
Relativity
Einstein's general theory, 4–5, 31–2,
67–8
Einstein's special theory, 4, 28, 66–7
Web sites, 145
A Relativity for the Layman (Coleman),
123
*Report on the Relativity Theory of
Gravitation* (Eddington), 32
Resonance linear accelerator, 34
Röntgen, Wilhelm Konrad, 6
biographical sketch, 89
Web site, 143
Royal Society of London, 121
Rutherford Appleton Laboratory (RAL),
121
Rutherford, Ernest, 8, *9*, 25, 26, 29, 30,
32
biographical sketch, 90
Web sites, 143

Sakharov, Andrei, Web site, 143
Salam, Abdus, 17, *91*
biographical sketch, 90–1
Scanning tunnelling electron
microscope, 44, 45
Schrieffer, John Robert, 13
biographical sketch, 92
Schrödinger, Erwin, 11
biographical sketch, 92–3
Web sites, 143–4
Schrödinger wave function, 11, 12
Schrödinger's Kittens (Gribbin), 125
Schwinger, Julian Seymour, 15
biographical sketch, 93

Seaborgium, 43
Segrè, Emilio Gino, *94*
biographical sketch, 93–4
Semiconductor diodes, tunnelling, 68
Shape-memory effect, 36
Shockley, William Bradford, 38, *39*
biographical sketch, 95
Sin-Itiro Tomonaga, 15
Soddy, Frederick, 25, 30, 32
Solar energy, 35
'Space-time', 4
Special theory of relativity, 4, 28, 66–7
Spectrum, 2–3, 5
Spin, 11
Standard model, 16
Static Electricity Web site, 144
Stephen Hawking's Universe Web site,
144
Strange Beauty (Johnson), 125
'Strangeness' in subatomic particles,
40, 74
String theory, 24
Strong nuclear force, 14, 17
Subatomic Logic Web site, 144
Subatomic particles, 14, 16
Eightfold Way classification scheme,
41
Gauge Bosons, 16
Hadrons, 16
Leptons, 16
'strangeness', 40, 74
Sun, total eclipse, *33*
Superconductivity, 13–14, 30, 36, 40–1,
45
high temperature, 86–7
Josephson effect, 41
work of John Bardeen, 52–3
work of Johannes Bednorz, 53
work of Leon Cooper, 62–3
work of Brian Josephson, 80
work of K. Alexander Müller, 86
work of John Schreiffer, 92
Superheterodyne principle, 25
Superstring: A theory of everything?
(Davies and Brown), 124
Surely You're Joking, Mr Feynman!
(Feynman), 71
Symmetry in nuclear physics, 100
Synchrocyclotrons, 37–8
Synthetic diamonds, 58, 59
Szilard, Leo, 18
biographical sketch, 95–6
Web site, 138
Szilard-Chambers reaction, 96

Tandem electrostatic accelerator, 50
Tau lepton (tauon), 43
Tau neutron, 46

Television, 7, 26
Teller, Edward, Web site, 131
Temperature Web site, 133
Tesla, Nikola, Web site, 144
The Theory of Electrons, 29
Thermodynamics
  first law, 33
  third law, 28
Thomson, J(oseph) J(ohn), 2, 9, 29, 31
  biographical sketch, 96–7
Time dilation, 4, 5
Time and space, 4–5
Ting, Samuel Chao Chung, biographical
  sketch, 97
Tokamak, 19
Transistors, 38, *39*, 41
  work of John Bardeen, 52
  work of William Shockley, 94
Transmutation of elements, 33, 37
Transuranic elements, work of Edwin
  McMillan, 83
Treatise on Radioactivity (Curie), 29, 63
Tributes to Paul Dirac (Taylor), 127
Triode valve, work of Lee de Forest,
  64–5
TRIUMF, 121–2
Tunnelling
  of electrons, 40
  in semiconductor diodes, 68

'U' subatomic particles, 43
Ultraviolet (UV) radiation, 6–7
Uncertainty principle, 12, 34
  work of Werner Heisenberg, 78–9
Unified electroweak theory, work of
  Steven Weinberg, 96–7
Unit converter Web site, 145
Unitary Field Theory (Einstein), 35
Universe, expansion, 23, 46
University of Cambridge Institute of
  Astronomy, 122
Unnilenium (Meitnerium), 44
Ununbium, 46
Unununium, 46
Upsilon meson, 43
Uranium, 7–8, 17, 19, 37
  fission, 18, 71–2

Villard, Paul Ulrich, 25
Visual Physics Web site, 145
VIXEN radar system, 50

'W' subatomic particles, 44
Walton, Ernest, 15, 97–8, *98*
  biographical sketch, 98
Wave mechanics, 34, 92
Wave nature of electrons, 64
Wave-particle duality, 59
Waves, Web sites, 137, 139
Weak nuclear force, 14, 17
Webster's Online Guide to Physics
  Web site, 145
Weinberg, Steven, 17
  biographical sketch, 98–100
Weinberg–Salam theory, 76
*What do you care what other people
  think?* (Feynman), 71
Wigner effect, 101
Wigner, Eugene P, 4, 18
  biographical sketch, 100
Wigner nuclides, 100
Wilson cloud chamber, 55
WIMPS (weakly interacting massive
  particles), 23
Women, contribution to physics,
  Web site, 130

X-ray crystallography, 6, 30, 31
X-ray spectroscopy, 81
X-ray tube, 32
X-rays, 28, 31
  and crystals, 82
  particle nature, 6
  work of Charles Barkla, 53
  work of William and Lawrence
    Bragg, 57–8
  work of Arthur Compton,
    60, 62
  work of Wilhelm Röntgen,
    89–90
Xenon, metallic, 43

Yukawa, Hideki, 14
  biographical sketch, 101

'Z' subatomic particles, 44
Zeeman effect, 83
Zeeman, Pieter, *3*
'Zeta' subatomic particle,
  44
Zweig, George, 16
  biographical sketch, 101
Zworykin, Vladimir, 7